Advance Praise for ***China and the Developing World***

"The ascension of China may be the most important geopolitical development during the first half of the twenty-first century. Yet it is also one of the least understood, given the uncertainties associated with China's capabilities, and, even more so, its intentions. *China and the Developing World* brings us much closer to understanding what is happening on our watch. The data are comprehensive, the analysis is sharp, and the insights are profound. Given the magnitude and the scope of the challenge China presents, it is especially gratifying to see a book of this quality coming primarily from a younger generation of scholars, as it will be their challenge to manage for the next several decades."

—**Richard Armitage**, former United States Deputy Secretary of State

"In an era of rising global concern over China's newfound national potency, precious little attention has been paid to Beijing's rapidly changing relations with the developing world. This timely new volume helps to fill that void. Bringing together contributions from nine experts, the book documents China's growing use of 'soft power' to win friends in the key emerging regions of South and Southeast Asia, Africa, the Middle East, and Latin America. Expertly edited, the volume offers important insights into the shifting interregional dynamics of China's search for global 'peace and security.'"

—**Richard Baum**, University of California at Los Angeles

"Publication of *China and the Developing World* represents an important turning point in understanding the purpose, scope, and implications of China's rise in Asian and world affairs. It provides the first comprehensive treatment of this important topic since the later stages of the cold war. Its clearly presented and balanced review and assessment of recent Chinese actions in what used to be called the third world will inform readers of the reality of Chinese foreign policy without hyperbole and with a discerning eye regarding the intentions and implications. It sets a foundation for further study of China's evolving international approach."

—**Robert G. Sutter**, School of Foreign Service, Georgetown University

"From Latin America to Africa, China is having a new and unexpected effect on governance, development, and security. This will require that policymakers and scholars improve their understanding of China's strategy in the developing world and knowledge of what commercial, diplomatic, and military interests motivate Beijing. In *China and the Developing World*, Eisenman, Heginbotham, and Mitchell have put together the first comprehensive assessment of China's rapidly growing relationship with the developing world. It is a must-read for anyone trying to make sense of China's sudden and important impact around the globe."

—**Michael J. Green**, School of Foreign Service, Georgetown University; former Senior Director for Asian Affairs, National Security Council

Reviews of *China and the Developing World*

"In *China and the Developing World*, Joshua Eisenman, Eric Heginbotham and Derek Mitchell cogently explain why a more confident leadership has chosen to demonstrate its muscle—and its appeal—in parts of the developing world. Many recent books on China's foreign policy simply list actions without attempting to divine a strategy. Eisenman, Heginbotham and Mitchell try to understand the reasons China's leadership makes decisions, though they know that the country's foreign policy remains a work in progress."

—London Review of Books

"This valuable collection of readings on China's relationship with the developing world . . . is timely and innovative."

—The China Quarterly

"Editors Joshua Eisenman, Eric Heginbotham, and Derek Mitchell, and a range of authors, offer the first in-depth examination of China's policies toward Central Asia, Africa, Latin America, South Asia, the Middle East, and Southeast Asia."

—China Digital Times

"Provides a good summary of information from sources otherwise hard to access—from Chinese government documents to African radio broadcasts—and paints a coherent bird's-eye picture of China's global presence."

—The China Journal

CHINA AND THE DEVELOPING WORLD

BEIJING'S STRATEGY for the TWENTY-FIRST CENTURY

JOSHUA EISENMAN, ERIC HEGINBOTHAM, and DEREK MITCHELL, Editors

Foreword by Kurt Campbell

KNOWLEDGE WORLD

KW Publishers Pvt Ltd

New Delhi

Joshua Eisenman, Eric Heginbotham, and Derek Mitchell, editors, *China and the Developing World: Beijing's Strategy for the Twenty-First Century* (Armonk, NY: M. E. Sharpe, 2007)

2010 First Indian Reprint.

ISBN 13: 978-93-80502-15-1

Published in India by

KW Publishers Pvt Ltd
NEW DELHI: 4676/21, First Floor, Ansari Road, Daryaganj, New Delhi 110002
T: +91.11.23263498 / 43528107 **E:** mail@kwpub.in / knowledgeworld@vsnl.net

MUMBAI: 15 Jay Kay Industrial Estate, Linking Road Extn., Santacruz (W), Mumbai 400054
T: +91.22.26614110 / 26613505 **E:** mumbai@kwpub.in
www.kwpub.in

Printed and Bound in India.

Contents

III. Conclusions

Appendices

Illustrations and Tables

Maps

Tables

Figure

Foreword

It has been nearly two decades since there has been any serious discussion of strategic competition among the great powers for positions of influence in what was then called the Third World. Yet, for much of the Cold War the United States and the Soviet Union—with China playing a bit part in East Africa and elsewhere during a particularly vigilant period of ideological intensity in the 1960s and 1970s—struggled for dubious advantages in forgotten ramparts in East Asia, Latin America, the Middle East, and Africa. Places like the Ogaden in Ethiopia, Jambo in southern Angola, and the mountainous jungles of Nicaragua all saw conflict and competition between Moscow and Washington as part of a seemingly inexorable global game of dominos. After the conclusion of the Cold War, it was widely hoped that the major powers would have an opportunity to shift their focus of attention from military confrontations in the developing world to more beneficial pursuits, including the alleviation of poverty, the promotion of exports and investments, and the general advancement of a multifaceted agenda to advance prosperity.

Alas, it has not turned out that way. What some thought might be a renaissance in relations between North and South looks conspicuously like strategic neglect—particularly for Africa—as many parts of the underdeveloped or slowly developing world struggle for attention and resources and attention from the United States and the other industrialized democracies. In recent years, China stepped into this generally bleak picture and the potential strategic consequences are enormous—not least of which is simply reminding others that large numbers of countries exist in forsaken places.

China participated in the earlier era of strategic competition for a time—building trains and providing military support and ideological sustenance to radical regimes across the globe—but withdrew to focus on domestic imperatives after 1978 and the beginning of economic reform. Now, decades later and with a much more advanced commercial status, but a largely undefined ideological and political agenda beyond its borders, China is again reaching out to either forsaken or largely forgotten areas of Africa, Latin America, and the Middle East.

Although China's relations with the developing world are a growing part of its larger foreign policy strategy of "peace and development," these new policy directions continue to receive surprisingly little attention in either academic or policy circles. Most books and essays on Chinese foreign policy deal with the topic only peripherally, as part of works on China's relations with its neighbors in the region and the West. This book, edited by Joshua Eisenman, Eric Heginbotham, and Derek Mitchell, fills an important gap. Its focus on China's strategy toward the developing world and its analysis of trends will aid academics and policymakers in their understanding of a topic that is gaining in importance and is already having implications for U.S. foreign policy.

This is the first full-length treatment of this subject in nearly two decades, since Dr. Lillian Craig Harris' publications *China's Foreign Policy Toward the Third World* (1985) and *China and the Third World: Champion or Challenger?* (1986). Much has changed, both within and beyond China's borders, between the appearance of Harris' books and the compilation of this edited volume. The transformation of geopolitics brought on by the collapse of the Soviet Union, Beijing's increasing press for unification with Taiwan, and China's own domestic economic and social reforms have led to a reevaluation of its overall strategies and objectives.

The last twenty years have seen enormous shifts in Beijing's foreign policy objectives and methods. While in the past, ideological and military concerns dominated Beijing's foreign policy aims in the developing world, the focus of Chinese foreign policy today is mainly business. China's furious pace of economic development has shaped a strategy in the developing world based on the country's need for energy, markets, and political support in the international community. Its growing acceptance of free market principles and a willingness to actively participate in regional and global associations, such as the World Trade Organization (WTO) and ASEAN+3, have helped integrate China into the larger global economic and diplomatic order. China has determined that in most circumstances—and at least for now—its needs are best met by seeking to shape the current global framework from inside the tent. This trend has generally added relevance and predictability to our current multilateral frameworks, but it has also ushered in a level of insecurity among members unsure of China's authoritarian system and future national intentions.

Beijing's overall objectives are best understood by focusing on its regional initiatives and bilateral relationships. Rather than a single-minded approach toward the developing world, Beijing's policies are now functionally organized. It has market policies, energy access policies, Taiwan policies, and policies designed to promote multipolarity. The pursuit of objectives in each of those areas is increasingly coordinated through regional initiatives, like the Shanghai Cooperation Organization and the China–Africa Cooperation

Forum, which provide opportunities to introduce region-specific policies to often overlooked areas.

Yet, despite China's changing aspirations, the authors correctly emphasize that significant continuity with twentieth century objectives endures. Since the Bandung Conference of 1954, China has consistently identified itself with the developing world, referring to itself as the "world's largest developing country" and making constant reference to shared historical experiences with the states of Asia, Africa, and Latin America. The acceptance of Beijing's claims to Taiwan also remains a condition for any developing country looking to establish official diplomatic ties with Beijing. China also continues to view the support of developing states as critical to its efforts in the international arena, particularly at the United Nations.

Perhaps the most important observation to come out of this book is that China's strategy toward the developing world remains largely a work in progress. Tensions between different priorities are already apparent, and these are likely to grow more discordant over time. For instance, does China see the developing world as an asset in its own struggles with the West, or will the United States and China eventually have generally reinforcing priorities? Is it ultimately about commercial opportunities or will strategic considerations grow in importance? How will China's foreign policy and national priorities be regarded once it graduates from being "one of the club" to being a great power in its own right? How will China's desire to reassert territorial and historical claims in Taiwan and the South China Sea be reconciled with its interests in further developing stable diplomatic and economic ties with Southeast Asia and elsewhere?

China faces difficult choices, and how it decides between these competing priorities will determine its role in the future world order. As China's economic, diplomatic, and military strength grows, so will its capacity to act on the world stage—including in areas of the developing world that were once thought to be too far afield. Through a detailed examination of Chinese strategic thinking and region-specific analysis, this volume provides the reader with a valuable prism through which to view the impact of China's rise in the developing world, as well as the repercussions for its neighbors and the United States. It is an important contribution to the literature on the meaning of China's rise and its implications for the world.

With Chinese power growing rapidly, and its leaders poised on the cusp of critical decisions about how to use that power, this book could not come at a more opportune moment. Certainly, each author brings a critical perspective and experience, but issues such as trade, energy, and the maintenance of support in the international arena cut across all the chapters. The result is a work that clarifies the various methods employed by Beijing to further its strategic objectives, as well as to develop those objectives. It is through this struggle to define national goals that China will define its larger place in the world, and the developing world will serve as a powerful case study for exploring the motivations of Beijing in a heretofore forgotten arena. This volume will serve a critical function for academics, commentators, and policymakers in their search to understand China's growing role in the world—both developed and developing.

Kurt M. Campbell
Senior Vice President, Kissinger Chair,
Director of International Security
Center for Strategic and International Studies

Introduction

Joshua Eisenman, Eric Heginbotham, and Derek Mitchell

China likes to call itself the world's "largest developing country." Yet, given its growing economic and political clout, China is hardly a typical member of the developing world. Even as China's remarkable rise remains incomplete and uncertain, its influence on global affairs is growing, and its actions to safeguard its interests are increasingly affecting the interests of the international community as a whole—whether Beijing recognizes or intends such a result or not. While China's relations with major powers such as the United States, Russia, Japan, and Europe have received due attention in recent years, Beijing's growing engagement with nations in the less developed "South" has only recently been noticed or assessed for its implications. In fact, China's growing engagement with these countries may have a substantial impact on the international system and global stability in coming years.

Beijing's relations with the developing world have seen a renaissance in recent years. China has concluded a series of notable agreements for energy and other resources with developing states in Africa, Latin America, the Middle East, and Asia. Chinese businesses have awoken to the trade, investment, and market opportunities of these less-developed regions. Chinese leaders are spending more and more time visiting and hosting counterparts from these regions to build personal relationships and discuss cooperation in areas of mutual interest. Government-sponsored infrastructure development, language schools, educational exchanges, and other forms of aid and assistance have increased China's "soft power" influence in many areas.[1] Eager to demonstrate its commitment to multilateralism and reduce concerns about its growing strength, Beijing now engages energetically in regional fora and has even created some of its own. Meanwhile, China's leaders have also continued to cultivate bilateral ties, signing formal "strategic partnerships" and bettering military-to-military ties with key states throughout the developing world.

While these developments are attracting increasing attention, they raise a number of analytical questions that are only just now being addressed. These questions range from how Beijing defines its interests and goals in the developing world, to how it has adjusted its tactics to achieve those aims, to what impact its new outreach is likely to have on global and regional dynamics, and U.S. interests.

China's Global Rise

Western observers have predicted for centuries that the "sleeping dragon" would someday wake to become a major political and economic force in regional and world affairs. Nonetheless, the pace and scope of China's emergence over the past generation has been nothing less than breathtaking. Between 1979 and 2005, according to Beijing's own figures, China's GDP increased from less than $150 billion to $1.65 trillion; its foreign trade climbed from $20.6 billion to $1.15 trillion; per capita income rose from $190 to more than $1200; and its share of the global economy grew from about 1 percent to nearly 4 percent.[2] Even the most optimistic policymaker in Beijing could not have predicted such a development when China began its economic "opening" in 1979 after several decades of ideological governance, and more than a century of domestic political turmoil, destructive division, and war that devastated the Chinese economy and society.

Indeed, over the last two decades, the Chinese Communist Party has come to accept that communism as an ideology is an obstacle to national development. Instead, the Chinese leadership today talks about developing China's "comprehensive national power," which includes a full array of economic, political, military, social, and cultural components; promoting a "harmonious society" at home; and restoring the nation's unity and territorial integrity, which Beijing defines largely in terms of Taiwan unification. With the collapse of communism as a unifying national ideology, the leadership's need to produce practical results in the pursuit of these goals and a better quality of life for its people take on added urgency to legitimize its rule.

The implications of this urgent shift toward domestic development and political and economic pragmatism are reflected in Beijing's apparent near-term foreign policy objectives: to secure a peaceful international environment, particularly on China's periphery; promote domestic trade and investment; isolate Taiwan as an international political entity to prevent its permanent separation; and build positive international relationships, particularly with the United States, to mitigate any unnecessary distractions or obstructions to China's handling of its domestic transition. In 2002, Chinese leaders began to refer to a period of "strategic opportunity," during which China will be free from external challenges to allow focus on internal development. In outlining this notion, Foreign Minister Tang Jiaxuan specifically noted that Beijing will seek the assistance of developed and developing countries alike to assist in China's situation, presumably so that the leadership may avoid the kind of social problems and economic turmoil that traditionally challenged the rule of Chinese Imperial dynasties for centuries.[3]

The Developing World

This domestic context is critical when considering China's contemporary relations with the developing world. While China viewed the Third World largely through an ideological prism during the Cold War (albeit with competitive geopolitical overtones related to both the Soviet Union and the United States), today China is reengaging these regions for highly practical reasons—primarily to find new markets for its goods and to fuel its growing economy's thirst for natural resources and energy supplies to power its industries and promote its growth. China consumes roughly one-third of total global output of steel, 40 percent of cement, and 26 percent of the world's copper, driving up global commodity prices and leading to fears of scarcity and resource competition. China is now the second largest global consumer of oil, with its overall share of world energy consumption rising from 9 percent to 12 percent in a decade.[4] Because available natural resources and raw materials are largely located in Latin America, Africa, and Asia, Beijing has placed increasing attention in recent years on its relations with nations in these regions.

As a result of these factors, China's total trade with the developing world grew 88 percent faster than its trade with the developed states between 1999 and 2003.[5] To facilitate this growing trade and make it sustainable in the face of trade frictions and local concerns about low-wage Chinese competition, Beijing has sought to deepen its political relationships with these states. It has established preferential duties on many exports from key trade partners in the developing world, and emphasized their common international interests both in bilateral and multilateral fora.

Strategic considerations have also helped propel Beijing's diplomacy in these regions. In Southeast Asia and Central Asia, for example, concerns about Beijing's long-term political and military direction reached alarming levels during the mid-1990s. Fearing isolation by a combination of local states with Washington, and sensing an opportunity to make political gains after the East Asian Financial Crisis in 1997–98, Beijing launched a series of diplomatic initiatives. It outlined a new security concept that emphasized cooperative security and confidence building. It invigorated its involvement in regional multilateral institutions, and worked to resolve longstanding territorial and boundary disputes, particularly along its continental land borders. And on the world stage, China's participation in UN peacekeeping operations, and a more proactive diplomacy with neighboring states (in part to cultivate its soft power), have succeeded in improving China's global image.

At the same time, China's relations with the developing world have sought to defend other important strategic national interests. Although China managed to strip away most of the key states recognizing the Republic of China during the 1990s, Beijing's competition with its "wayward province"—played out largely in Latin America and Africa—remains intense. To both avoid censure in the United Nations, and to curry favor with resource-rich states, China has helped to shield a variety of unsavory regimes from international action over their human rights records and provided economic and other assistance to bolster their hold on power. Beijing has often formed strategic partnerships and other relationships with states that share an interest in fostering a more "multi-polar" international environment, in apparent counterpoint to American power.

Indeed, the United States and other nations are watching closely how Beijing uses its growing influence in the developing world as an indication of what kind of power China will be as it grows—whether it will be, in the words of one senior U.S. official, a "responsible stakeholder."[6] China's leaders often reassure the international community that their intentions are benign. Some analysts in the West, in fact, believe that China's integration into the global economy will serve as a powerful constraint on China's international behavior as it provides the state with a greater stake in the health and stability of the international system. Others worry that China's unique focus on its own economic needs, and the stakes of failure for the Chinese Communist Party leadership should it fail to meet those needs, will produce a purely self-interested foreign policy that will engage unconditionally with unsavory nations and thereby challenge the efforts of the international community to uphold established norms of the international system, such as non-proliferation, human rights, environmentalism, and clean governance. This volume will evaluate the nature and character of China's relationships in the developing world today for trends in China's international behavior and assess their implications on global affairs.

Existing Literature

While the analytical literature on Chinese relations with the West has grown over the last decade, work on Beijing's interests in and policies toward the developing world remains sparse—despite the importance of the topic both for intrinsic reasons and for the signals it might provide about China's overall posture and direction.

The decline in attention over the years, from a historical standpoint, is understandable. During the Cold War, when the United States saw itself as engaged in a global struggle against the spread of communism, China's interests in the developing world—and support for "national liberation movements" during the 1960s and 1970s—were well covered in Western surveys of Chinese foreign policy.[7] As China focused inward, and its role in the developing world was perceived as less relevant to the United States, Western literature and expertise on China's relationship with the developing world diminished in the 1980s.[8]

Much of the contemporary literature on Chinese foreign policy is framed by a vigorous debate between relative optimists and pessimists. Optimists see little immediate cause for concern regarding China's present foreign policy, and a number of positive signs regarding recent trends in its diplomacy. They note that Beijing's freedom of maneuver is severely constrained by its domestic weaknesses, dependence on foreign investment, and membership in international organizations.[9] And they argue that the worldview of China's elite policymakers—and even more that of its bureaucratic actors—is evolving as they internalize the values associated with international organizations and their own constructive rhetoric.[10] Optimists note that the elements of China's "new diplomacy" include a greater emphasis on engagement with multilateral institutions, confidence building, resolution of territorial disputes, and improved (if still rudimentary) transparency.[11]

On the other side are those who take a dimmer, or more skeptical, view of Chinese behavior. Some on this side consider an authoritarian China to be implacably hostile to the United States, and antithetical to U.S. interests and values. Other pessimists argue that regime type matters little, but that China (as a rising power) and the United States (as the preeminent global power) are nevertheless fated to clash over the shape and rules of the international system.[12] They tend to view United States-China relations as a zero-sum game, in which China's rise presents an inevitable threat to U.S. supremacy and global power. They contend that although China may be constrained by weakness today, its behavior is likely to change as it grows stronger in the future. The more radical of the pessimists believe there are few limitations on China's future ambitions.[13]

A third group contends that conflict, whether between China and the United States or between Beijing and one or more of its neighbors, is more likely to occur as a result of more specific or particular differences, compounded by domestic politics, nationalism, and the types of miscues that frequently trouble international relations.[14] Taiwan is regarded as a particularly likely source or focus of conflict, given the complex and unstable mix of domestic and international variables in play there.

In short, while China's reinvigoration of its relationships in the developing world has led to a flood of media reports and studies in recent years, there has been little effort to place these recent developments in the broader context of

China's overall foreign policy and history of relationships with those regions. This volume represents the first book-length treatment of China's role in the developing world since the end of the Cold War.[15]

Framework/Questions

This volume adopts a comparative approach to examining patterns and implications in China's relations with the developing world. Each of the core chapters examines Beijing's overall regional relations, its relations with individual regional states, the interests underlying these relationships, and Beijing's regional strategy. Each contributor provides a brief historical background of the PRC's involvement in the region and an assessment of changes over time, with a particular focus on developments since the reorientation of Beijing's foreign policy beginning in the late 1990s.

Each author has addressed four sets of questions:

First, how does China define its interests in the region, and has that definition changed over time? What policies does China pursue to support its declared interests and its larger foreign policy goals?

Second, what methods or tactics does Beijing employ with nations in the region to achieve its objectives? What assets does it bring to bear, and what weaknesses continue to hamper its efforts?

Third, how successful has Chinese diplomacy been in these regions? Has Chinese influence grown and, if so, upon what does its influence rest? If not, what are the obstacles to Chinese influence?

Fourth, what are the implications of China's policies toward other actors? What impact does Beijing's new outreach have on regional affairs, including governance and the overall welfare of the region's citizens? What impact does Beijing's recent engagement have on U.S. interests? And what impact does it have on the preservation or evolution of international norms and institutions?

These questions are addressed in the chapters that follow. Derek Mitchell and Carola McGiffert open the volume with an historical overview of China's relations with the outside world during China's Imperial history and through the Cold War. They trace continuities and change in China's approach over the centuries, from its complex relationships with the world outside its cultural and territorial boundaries during its Imperial days, to Beijing's support for national liberation movements in the Third World during the Cold War, to the reawakening of interest in the developing world for less ideologically-driven economic and political reasons in recent years. Mitchell and McGiffert note that many echoes of China's traditional principles and approaches to dealing with lesser-developed areas along its border and beyond remain in its contemporary international outlook.

Joshua Eisenman examines China's extensive network of diplomatic and commercial ties in Africa. He notes that China's once strongly ideological foreign policy in Africa has become largely a quest for strategic and economic influence. Eisenman comments that the establishment of the China-Africa Cooperation Forum in 2000 has ushered in a new age in China's relations with African nations, and allowed Beijing to institutionalize a growing array of technical training, debt relief, financial aid, and infrastructure development projects. He notes that while China's economic assistance and investment generally contribute to the betterment of Africa, its arms sales and often heavily unbalanced trade relationships do not.

In chapter 3, Matthew Oresman describes China's growing relationship with Central Asia. Oresman notes China's interest in demonstrating the virtues of its "new security concept" of informal strategic partnerships over formal alliances, and acquiring necessary resources for its economic development. Oresman discusses how each of these elements connect closely to China's general goal of "strategic denial": to deny destabilizing influences the ability to cross the border to challenge China's internal security; deny the use of Central Asia by the United States to contain China; and deny Russia from monopolizing influence along China's border. In particular, Oresman examines the importance—and limitations—of the establishment and development of the Shanghai Cooperation Organization to help China, Russia, and the Central Asian nations build confidence and address the three evils of terrorism, separatism, and extremism.

Chung-chian Teng has conducted a thorough review of China's fast expanding relations with the Latin American nations in chapter 4. Teng explores whether China will be a responsible power in Latin America or threaten the existing order and challenge U.S. predominance in the region. Latin American natural resources, agricultural products, and markets help allow China meet its domestic needs; at the same time China's purchases and investments aid regional development, he concludes.

Mao Yufeng's chapter 5 draws attention to increasing Chinese activism and influence in the Middle East and the Arab world. Because of its increasing dependency on oil imports from the region, Beijing is mobilizing hard and soft power resources to promote economic, political, and cultural relations with the Middle Eastern countries. Mao writes that Beijing is strategically avoiding confrontation with the United States in that region, while gradually seeking to expand its own influence there.

In chapter 6, Rollie Lal addresses the increasingly close, but still uncertain, relationship between China and India. She begins by highlighting the historical distrust that has long kept these two emerging great powers at arms length. She argues that economic interests are now pulling New Delhi and Beijing closer together. Although both China and India are developing economies, Lal observes that they are largely complementary. In the Chinese case, a desire to prevent India from being seduced into an anti-China coalition by the United States provides an additional powerful motivation. She concludes by both acknowledging the progress both have made toward building a "strategic partnership," while cautioning that there are still many tensions and uncertainties in the relationship.

Michael Glosny discusses how China has made a concerted effort to reassure Southeast Asia about its constructive intent and benign orientation in chapter 7. He comments that the region has generally welcomed this effort and benefited from China's emergence in recent years, although the nations remain wary about the full implication of growing Chinese strength. Glosny notes that China's diplomacy in the region has not pushed any of these countries to act contrary to their interests, or to side with China against the United States, but instead China seeks to gain access to critical natural resources, develop new markets for its products, prevent the development of an anti-China coalition along its border, and stabilize its periphery to allow Beijing to focus on its internal challenges.

Finally, Eric Heginbotham weaves together the findings of the individual chapters into a concluding analysis of themes in China's engagement with the developing world in chapter 8. He outlines the pressures that led Beijing to reevaluate its foreign policy by the late 1990s, and reviews the elements of its more proactive "new diplomacy." Heginbotham then evaluates the question of China's influence in the developing world, arguing that while its influence has expanded as the nation's economic weight has grown, that influence is largely confined to areas that are most relevant to its defensive, or counter-containment objectives. Finally, he concludes with an assessment of the impact of China's new diplomacy in the developing world on global and regional stability, U.S. interests, and on emerging international norms.

One may measure and analyze China's remarkable rise over the past generation, but predictions of where the Middle Kingdom is going and its effect on the international system must be made with great modesty. Among much uncertainty, however, the one certainty is that China will be both a challenge and an opportunity for the international community that will require close scrutiny, careful management, and sober understanding in the coming years.

Notes

1. Professor Joseph Nye coined the phrase "soft power" in the early 1990s to describe, as he put it, "the ability to get what you want through attraction, rather

than coercion or payments." Alternatively, Nye defined the term as "shap[ing] the preferences of others" to do things in your interest through the attractiveness of one's culture, political ideals, and policies, and leading by example. See Joseph S. Nye, Jr., *Soft Power: The Means To Success in World Politics* (New York: Public Affairs, 2004), x, 5.

2. Speech by State Councilor Tang Jiaxuan, "Vigorously Promoting China-U.S. Constructive and Cooperative Relations in the New Century," address at Welcome Luncheon by National Committee on U.S.-China Relations and U.S.-China Business Council, Washington D.C., July 27, 2005.

3. For a comprehensive overview of China's outlook on the international security environment, and on Chinese foreign and security strategy as outlined by then-Foreign Minister Tang Jiaxuan, see "Chinese FM: China's Int'l Status Grown in Last 13 Years," Statement at 16th National Congress of the Communist Party of China, Xinhua News Agency, October 9, 2002.

4. Nightly Business Report, February 22, 2005.

5. IMF, *Direction of Trade Statistics*.

6. "Whither China: From Membership to Responsibility?" remarks of Deputy Secretary of State Robert B. Zoellick, National Committee on U.S.-China Relations, September 21, 2005.

7. Since China's role in the Third World dominated Chinese foreign policy for much of the Cold War, the topic receives substantial coverage in most treatments of the latter during this period. See, for example, Jerome Alan Cohen, *The Dynamics of China's Foreign Relations* (Cambridge: Harvard University Press, 1970); Shen-Yu Dai, *China, the Superpowers, and the Third World: A Handbook on Comparative World Politics* (Hong Kong: Chinese University of Hong Kong Press, 1974); Alvin Z. Rubinstein, *Soviet and Chinese Influence in the Third World* (New York: Praeger, 1975); Richard Lowenthal, *Model or Ally? The Communist Powers and the Developing Countries* (New York: Oxford University Press, 1977); James C. Hsiung and Samuel S. Kim, eds., *China in the Global Community* (New York: Praeger, 1980); G.W. Choudhury, *China in World Affairs: The Foreign Policy of the PRC Since 1970* (Boulder: Westview Press, 1982); and King C. Chen, *The Foreign Policy of China* (Roseland, NJ: East West Who, 1972). Surveys of China's relations with specific regions include: John C. Cooley, *East Wind Over Africa: Red China's African Offensive* (New York: Walker and Company, 1965); Emmanual John Hevi, *The Dragon's Embrace: The Chinese Communists in Africa* (New York: Praeger, 1966); Alaba Ogunsanwo, *China's Policy in Africa 1958–1971* (Cambridge: Cambridge University Press, 1974); Bruce D. Larkin, *China and Africa 1949–1970: The Foreign Policy of the People's Republic of China* (Berkeley: University of California Press, 1971); Cecil Johnson, *Communist China and Latin America 1959–1967* (New York: Columbia University Press, 1970); Yitzhak Shichor, *The Middle East in China's Foreign Policy* (New York: Cambridge University Press, 1979); Melvin Gurtov, *China and Southeast Asia—The Politics of Survival: A Study of Foreign Policy Interaction* (Baltimore: Johns Hopkins University Press, 1975); and Joseph Camilleri, *Southeast Asia in China's Foreign Policy* (Singapore : Institute of Southeast Asian Studies, 1975).

8. The exception that proved the rule were two books involving Lillian Craig Harris: Lillian Craig Harris, *China's Foreign Policy Toward the Third World* (New York: Praeger, 1985); and Lillian Craig Harris and Robert L. Worden, eds., *China*

and the Third World: Champion or Challenger? (Dover: Auburn House, 1986). A search of JSTOR, an online search engine for academic journals, shows a dramatic decline in, for example, the number of articles including mention of "China" and "Africa" over each of the last two full decades compared to the one before. Declines were recorded for articles on China's relations with Latin America and the Middle East. The declines were registered both in absolute numbers and relative to the total number of articles included in the database. Only articles published in and before 1999 are included, so more recent trends cannot be assessed from that source.

9. On China's economic vulnerabilities and its continuing dependence on openness, see George J. Gilboy, "The Myth Behind China's Miracle," *Foreign Affairs*, July/August 2004; and Nicholas R. Lardy, *Integrating China Into the Global Economy* (Washington, D.C.: Brookings Institution Press, 2002). On social weaknesses and instability, see Murray Scot Tanner, "China Rethinks Unrest," *Washington Quarterly*, summer 2004. On how economic and social weakness constrains behavior, see Phillip C. Saunders and Erica Strecker Downs, "Legitimacy and the Limits of Nationalism: China and the Diaoyu Islands," *International Security,* winter 1998/1999. While somewhat ambivalent about causality, a cautiously optimistic statement is provided in Alastair Iain Johnston, "Is China a Status Quo Power?" *International Security*, spring 2003.

10. Advocates of this view are cautious in drawing conclusions, but nevertheless see changing elite values. See, for example, Alastair I. Johnston, "Socialization in International Relations: The ASEAN Way and International Relations Theory," in G. John Ikenberry and Michael Mastanduno, eds., *International Relations Theory and the Asia-Pacific* (New York: Columbia University Press, 2003); Evan S. Medeiros, *Shaping Chinese Foreign Policy: The Evolution of Chinese Policies on WMD Nonproliferation and the Role of US Policy, 1980–2004*, unpublished book manuscript; and Elizabeth Economy, "The Impact of International Regimes on Chinese Foreign Policy-Making: Broadening Perspectives and Policies . . . But Only to a Point," in David M. Lampton, ed., *The Making of Chinese Foreign and Security Policy in the Reform Era* (Stanford: Stanford University Press, 2001).

11. Evan Medeiros and M. Taylor Fravel, "China Takes Off," *Foreign Affairs*, November/December 2003; David Shambaugh, "China Engages Asia: Reshaping the Regional Order," *International Security*, winter 2004/2005; and Rosemary Foot, "Chinese Power and the Idea of a Responsible State," *The China Journal*, January 2001. For more descriptive accounts see, for example, Jane Perlez, "The Charm from Beijing: China Strives to Keep Its Backyard Tranquil," *New York Times*, October 8, 2003; Amitav Acharya, "China's Charm Offensive in Southeast Asia," *International Herald Tribune*, November 8–9, 2003; and Philip Pan, "China's Improving Image Challenges U.S. in Asia," *Washington Post*, November 15, 2003.

12. For this "structural realist" perspective, see John J. Mearsheimer, *The Tragedy of Great Power Politics* (New York: W.W. Norton, 2001); Aaron L. Friedberg, "The Struggle for the Mastery of Asia," *Commentary*, November 2000; Aaron L. Friedberg, "Ripe for Rivalry: Prospects for Peace in a Multipolar Asia," *International Security*, winter 1993/94; and Ashley Tellis, "A Grand Chessboard," *Foreign Policy,* January/February 2005.

13. Richard Bernstein and Ross H. Munro, *The Coming Conflict with China* (New

York: Knopf, 1997); and Bill Gertz, *The China Threat: How the People's Republic Targets America* (Washington: DC: Regnery Publishing, 2000).

14. See Thomas J. Christensen, "Posing Problems Without Catching Up: China's Rise and Challenges for U.S. Security Policy," *International Security*, spring 2001.

15. The most recent works include Samuel S. Kim, *The Third World in Chinese World Policy* (Princeton: Princeton University, 1989), and the edited volume by Lillian Craig Harris and Robert Worden, eds., *China and the Third World: Champion or Challenger?* (Dover, MA: Auburn House, 1986).

I

Analytical Background

1

Expanding the "Strategic Periphery"

A History of China's Interaction with the Developing World

Derek Mitchell and Carola McGiffert

On his most famous voyage in 1492, Christopher Columbus led a fleet of 90 sailors aboard three ships, the longest of which was 85 feet. This historic journey was heralded as a breakthrough in maritime exploration. Yet a half century earlier, Zheng He, the Chinese Muslim eunuch, commanded 300 ships of up to 400 feet, equipped with the most advanced maritime technology of the day and manned by 28,000 sailors, on maritime exploits far from the shores of China. Zheng's sophisticated fleet dwarfed anything sailing out of Europe at the time, and during his seven historic voyages he spread Chinese goods and culture from Southeast Asia to India and on to the shores of Africa. China was at the height of its global power during the early years of the Ming Dynasty (1368–1644 CE), and Zheng He's voyages were a landmark in China's interest and outreach to the world around it.[1]

In fact, Zheng He's expeditions emerged out of an intermittent Chinese maritime tradition and rich interaction with its neighbors reaching back to the first millennium CE. It was China, after all, that invented the compass during the Han Dynasty (206 BCE–220 CE) and first applied the invention to seafaring during the Song era (960–1279 CE) when China established its first national navy as the most powerful and sophisticated of its day.[2] During the Han Dynasty, Buddhism was introduced to China via overland trade routes that extended to Central Asia and Persia. During the Tang Dynasty (618–907 CE), pilgrims from India and scholars and traders from China and Central Asia traveled along the Silk Road, over which trade in silks, spices, and tea reaching as far as Rome introduced commercial and cultural exchange to the seat of the emperor in Chang'an (Xi'an). The Tang period led to growing Chinese influence in Japan and Korea, whose people adopted the Chinese writing system and other cultural, artistic, and philosophical concepts. Even today, the Chinese look to the Han and Tang dynasties as a hallowed historic era during which China assumed its place as the acknowledged leader in regional, if not global, affairs.

Indeed, had it not been for a power struggle in which suspicious, xenophobic Confucian scholars prevailed in the late fifteenth century, it might have been Zheng He, rather than Columbus, who "discovered" America and officially established America's patrimony as Chinese rather than European. Instead, China's official engagement with the world beyond its immediate periphery ended soon after Zheng returned home, as China turned inward, dismantled its maritime forces, and inaugurated several centuries in which China would sink into self-imposed semi-isolation. China's international engagement returned to a traditional and highly regulated system of foreign interaction that fueled China's sense of self-importance, but led to complacency in science, commerce, and military affairs that would eventually lead to the tumultuous end of China's Imperial era.

Decades of humiliation, division, and war in the nineteenth and twentieth centuries led China to ponder its past and the traditional attitudes that allowed the Middle Kingdom to become weak and vulnerable to external challenges. Although sometimes considered the latest Chinese dynasty, the Chinese communists under Mao Zedong took a much different approach to China's security and to its relationship with the world. During the early years of the Cold War, pitting two ideologies and power blocs against one another, China chose to focus on the so-called "Third World" of developing nations to promote its own ideological and political agenda. By the 1980s and 1990s, with the decline of ideology and rise of economic reform and pragmatism, China's relationship with the developing world became subordinated to a priority interest in establishing constructive political and economic relationships with the developed world. This was done to promote domestic development and stability, and to return China to the ranks of the world's great powers.

Nonetheless, China today continues to call itself the world's largest developing nation, even as it reaches out and connects itself to the international community to a degree unprecedented in China's long history. The basis for China's approach to international affairs generally—and the developing world specifically—will likely be built upon the foundation of China's experience and history of dealing with the outside world. Examining the themes of China's external relations throughout its history can provide the necessary context for understanding China's current approach. This chapter seeks to provide such a context, and to assess what elements may—or may not—continue to pertain today as China emerges as a major global player.

While China has historically had substantial and productive interactions with peoples and cultures beyond its borders, Imperial China's external relations through the centuries nonetheless were infused more with alienation than outreach. Imperial China's policy and perspective toward the outside world drew greatly on its cultural heritage, unique geopolitical situation, and

Imperial Dynasties and Governments of China

- Xia Dynasty: 2206–1766 BCE
- Shang Dynasty: 1766–1122 BCE
- Zhou Dynasty: 1122–770 BCE
- Spring and Autumn Period: 770–476 BCE
- Warring States Period: 476–221 BCE
- Qin Dynasty: 221–206 BCE
- Han Dynasty: 206 BCE–220 CE
- Three Kingdoms Period: 220–265
- Jin Dynasty: 265–420
- Northern and Southern Dynasties: 420–581
- Sui Dynasty: 581–618
- Tang Dynasty: 618–907
- Five Dynasties and Ten Kingdoms: 907–960
- Song Dynasty: 960–1279
- Northern Song: 960–1127
- Southern Song: 1127–1279
- Liao (Khitan) Dynasty: 916–1125
- Jin (Jurchen) Dynasty: 1115–1234
- Yuan (Mongol) Dynasty: 1279–1368
- Ming Dynasty: 1368–1644
- Qing (Manchu) Dynasty: 1644–1911
- Republic of China: 1912–present (relocated to Taiwan in 1949)
- People's Republic of China: 1949–present

Note: For excellent comprehensive discussions of Imperial China's external relations, see *The Chinese World Order: Traditional China's Foreign Relations*, ed. John King Fairbank (Cambridge, MA: Harvard University Press, 1968); John King Fairbank and Edwin O. Reischauer, *China: Tradition and Transformation* (Boston: Houghton Mifflin Company, 1973); and Mark Mancall, *China at the Center: 300 Years of Foreign Policy* (New York: The Free Press, 1984). For an outstanding analysis that traces Chinese strategy from the Imperial period to the present, see Michael Swaine and Ashley Tellis, *Interpreting China's Grand Strategy: Past, Present, and Future* (Santa Monica, CA: RAND, 2000).

consistent requirement to attend to its endemic internal challenges. Specifically, China's adherence to Confucian philosophy, its proud self-image, its constant concern over the vulnerability of its periphery to external challenges, and its consistent fear of domestic turmoil—if not dissolution—due to internal corruption, civic unrest, or palace intrigue drove what was often an essentially defensive approach to external relations. During periods of strength, however, Chinese emperors often displayed more aggressive, expansionist tendencies. Each of these themes, and the interplay among them, will be examined below.

Self-Image and Cultural Values

It is commonly noted that during the Imperial period, China's view of the world and its place in it was reflected in the name the Chinese gave their country—*Zhongguo*, or "Middle Kingdom"—and in the term used to describe the scope of their imperial suzerainty—*tianxia*, or "all under heaven." China's advanced culture as well as material and philosophical achievements by early in the first millennium CE indeed led to a pronounced self-reverence that informed every aspect of the nation's behavior, from imperial rites and rituals to how the Chinese viewed themselves in relation to the outside world.

Those outside Chinese civilization were officially termed "barbarians," and thus unworthy of study for the purpose of foreign relations or otherwise. During the Ming Dynasty, for instance, Chinese descriptions of foreign lands portrayed foreigners in mystical and fantastical terms—akin to exotic animals—and described in deliberately condescending or belittling language.[3] It has also been said that during the Imperial era the Chinese viewed the entire world outside their borders as the developing world, with which the Chinese emperor, as the "Son of Heaven," might engage only with great condescension and charity because of their unfortunate status of not being a part of Chinese civilization.

The Chinese emperor and Confucian scholar/bureaucrat viewed the world as they viewed the organization of their own society, through a Confucian notion of hierarchy and reciprocal responsibility between ruler and subject, father and son, or older brother and younger brother. China put itself at the pinnacle of this world order, with its leader holding the self-proclaimed "mandate of heaven" as the basis of his legitimacy to rule.

The emperor's mandate was based on a moral authority conveyed by adherence to the Confucian value system. According to Confucian values, the international system—like the social system—should be led by a civilized force that exercises rule by virtue rather than by naked power.[4] Confucian bureaucrats tended to disdain the professional military as a matter of principle, viewing military power as beneath the dignity of an advanced civilization. Confucianists even declined to include warriors in their social and occupational hierarchy of scholar-bureaucrat, farmer, artisan, and merchant, in descending order of status.[5] In their dislike for military force, Confucians made an exception on occasion, particularly when an external power failed to adhere to Chinese superiority. Such a development required the Son of Heaven to firmly return the violator to its appropriate place in the Sino-centric order.[6]

China's superiority complex was facilitated by its lack of interaction with societies of equal cultural achievement, at least from China's perspective. Aside from interaction between Chinese traders and foreign counterparts,[7] most of China's foreign contact during the Imperial era was either with the relatively undeveloped militant nomadic and seminomadic cultures on China's northern and western borders, or with neighbors in southeast and northeast Asia whose cultures often were influenced by China. China's ritualized interaction with these border nations and tribes over the centuries, as discussed below, only served to reaffirm its sense of superiority and exceptionalism. And even when unsuccessful in preventing occupation by non-Han peoples, as during the Mongol (Yuan: 1279–1368 CE) and Manchu (Qing: 1644–1911 CE) dynasties, the eventual preservation of the Confucian order and cultural assimilation of their occupiers only further reaffirmed the superiority of the Chinese culture.

Imperial China was generally self-sufficient in terms of natural resources, which also enabled its rulers to feel little need to compete with other nations to sustain its power or legitimacy. What China might have received from the outside world tended to be viewed as luxuries or objects of curiosity rather than materials critical for the stability and sustainability of the national economy and polity.[8] Generally, Chinese leaders focused their energies and strategies internally to secure their rule, rather than externally to dominate foreign powers.

Superiority and Weakness: Imperial Chinese Relations with the Strategic Periphery

Despite China's confident sense of its own cultural superiority relative to the world around it, its leaders were quite aware of the country's vulnerability to challenges from its immediate inland periphery—particularly to the north and west (China tended to ignore its maritime borders as they were largely safe from foreign challenges until the middle of the Qing dynasty). The purpose was to safeguard the Chinese heartland, which was generally defined as encompassing the maritime eastern shore, north through the Yellow River plains and tributaries to today's Beijing, south through the Yangtze River region to the South China Sea, and as far west as the mountains and high plains of modern Sichuan and eastern Qinghai.[9]

Attention to challenges from the west began at least as far back as the first millennium, during the Han Dynasty, which suffered from attacks primarily from the Turkic Xiongnu (Hun) tribes. Safeguarding trade along the Silk Road required securing the areas along China's border with Central and South Asia. Invasions from the north led to the establishment of two major dynasties not run by the ethnic Han people during the second millennium—the Mongol (Yuan) and Manchu (Qing)—and several small dynasties that ruled over parts of China—the Qidan (Liao: 916–1125 CE) and the Jin (Jurchen/Ruzhen: 1115–1234 CE). The security of China's periphery, therefore, was a very practical, immediate, and strategic concern to Chinese leaders. Pacification of Inner Asia became consistent elements of Chinese Imperial strategy.

Chinese leaders handled the challenge from the border in several different ways, often depending on China's own internal situation. When China was beset by internal troubles, usually during a period of dynastic consolidation at the start of a new era, or during the late, turbulent years of a dynasty, Chinese leaders took an essentially defensive position to hold off border threats and challenges. When stronger, it took the fight to the "barbarians" and expanded its strategic periphery accordingly.

During the Han Dynasty, for instance, leaders sought to fight off, buy out, or pit barbarian against barbarian.[10] Ultimately at a military disadvantage, fighting far from the Chinese heartland, Han Dynasty emperors often resorted to a policy of "peace and kinship" by appeasing nomad leaders with lavish gifts, entertainment, and even betrothal to Han princesses.[11] Over time, such interactions became ritualized as a method of economic (trade) and diplomatic intercourse between China and non-(ethnic) Han peoples. Despite being established initially to deal with the essential vulnerability of the Chinese empire, the process acquired the veneer of Chinese superiority over time when the Confucian bureaucracy decided to frame the ritual exchanges as a barbarian "tribute" to the superior (virtuous) Chinese empire.[12] The visiting state was required to use highly specific and formalized language of subservience, present the Chinese court with gifts (sometimes including a prominent hostage), and perform the ritual kowtow—or "three kneelings and nine prostrations"—in a minutely choreographed display of fealty that, in the Chinese mind, provided the Middle Kingdom with a formal acknowledgment of its cultural and political superiority.[13] In return, the Chinese provided lavish gifts and allowed limited trade between the two sides.

China usually gave much more than it received from these missions, reflecting the court's attitude that its smaller neighbors had little to offer their great nation, and demonstrating Chinese generosity. In return for receiving Chinese political recognition and protection, China required reciprocal obligations from tributary states to assist in keeping the peace, including serving alongside Chinese armies whenever called upon.[14]

In the end, the system primarily served the practical purpose of facilitating trade for the tributary states, rather than incurring loyalty and obedience. The issues of relative status, superiority and inferiority, and reciprocal responsibility were generally of lesser importance to them. While flawed, the tributary system offered mutual benefit from both economic and security standpoints to the tributary states and China alike. Tributary states received trade benefits and, in some cases, security guarantees, while China got strategic peace of mind, reaffirmation of its self-regard, and an effective means of saving the cost of maintaining a large standing army to patrol all its borders.

Periods of Strength and Confidence: Imperial China Reaches Out

Once an Imperial dynasty consolidated its control within China and felt relatively strong and confident, however, a different China often emerged from the one that assumed a defensive posture during periods of vulnerability. This China sought to exert more direct control over its periphery, and even expand through military means its territorial—and tributary—boundaries.[15]

Vietnam and Korea were favorite targets. The Han Dynasty overran northern Vietnam, leading to Chinese rule down to modern Hanoi.[16] The Sui sought to extend Chinese dominion even farther, to southern Vietnam.[17] The Song failed in its campaign against its southern neighbor; and the Ming initially succeeded in its attempt at control but found the cost of trying to formally absorb Vietnam into the Chinese empire too high, resorting ultimately to a traditional tributary relationship.[18] Similarly, Korea experienced repeated Chinese efforts at direct control or subjugation. The Han, Sui, and Tang each attempted to use military means to absorb the peninsula (or parts thereof) into the Chinese imperium, until the Ming and Qing settled for a dominant tributary relationship that made Korea a virtual protectorate.[19]

The Ming Dynasty's founding emperors, still worried about the Mongol threat even after successfully casting off the Yuan Dynasty, took the offensive against nomadic threats from the north and northwest to extend their Confucian suzerainty even farther. Beginning in 1405, and for the next two decades, Ming Emperor Zhu Di sent Zheng He on at least seven maritime expeditions to forcefully secure new tributary relationships and economic benefits for China in the South Pacific, Indian Ocean, Persian Gulf, and the eastern coast of Africa[20]—including helping to overthrow the leaders of Sri Lanka and Sumatra and replace them with more tributary-friendly rulers.[21] Although private maritime and overland merchant trade to Southeast, Central, and South Asia had occurred for centuries, these official voyages to nearly fifty new countries constituted the most extensive official contact with the non-Confucian world in Chinese history to that point. While official China retreated into relative diplomatic isolation in the ensuing centuries, Zheng's voyages nonetheless set the stage for a dominant presence of Chinese traders in Southeast Asia.

Ming and Qing Dynasty leaders continued to pursue territorial expansion into regions previously considered outside the Chinese heartland. They did so through both military conquest and political cooptation. The founding Ming emperor, for instance, referred in an early proclamation to the "countries of Yun-nan and Japan,"[22] suggesting that he considered the former—which today is considered an uncontroversial province of China—an independent entity outside China's frontiers. Through a process of steady cooptation of local authorities by officials from the Ming bureaucracy, the formerly autonomous (or semiautonomous) entity became formally incorporated into the Chinese empire by the end of the Ming Dynasty.[23] Such a process of absorption was not unique to Yunnan or the Ming period, but it provides a window into one way that China was able to expand its territory steadily over the centuries continuing into the Qing Dynasty that followed.

Early Qing emperors also expended much effort in attempts to influence and control China's periphery beyond that achieved by earlier dynasties, including pacification efforts against Burma, Nepal, and Mongolia. In the process, Qing leaders aggressively incorporated Chinese Turkestan (Xinjiang today), Taiwan, and Tibet into its Imperial order as protectorates, which set the stage for later claims of explicit sovereign rights to these relatively autonomous regions by Chinese rulers in the twentieth century.[24]

The formation of the Chinese Imperial Empire through military action and gradual assimilation of surrounding peoples thus suggests another way to view China's posture toward the developing world along its border during this period. In fact, the Chinese imperium tended to expand and contract throughout history, making historic "China" a geographically difficult concept to define, and cross-border aggression difficult to identify. The Great Wall exists far inside the borders assumed by today's People's Republic of China. Today's Tibet, Xinjiang, Manchuria (Heilongjiang), Inner Mongolia, and Taiwan, among other areas, were historically considered by ancient China as dangerous sources of barbarian threats, and thus were hardly part of the imperium.

In the end, as China forcefully absorbed many of the peoples on its northern and western borders into its imperial orbit, the definition of China's strategic periphery necessarily expanded as well. As more and more territory became part of the Chinese imperium, more and more territory needed to be protected through an extended buffer. In the end, recognition from the empire of the object's inherent sovereign legitimacy was not often the determining factor in whether a border nation, people, or area maintained its sovereign independence and autonomy from the Chinese empire. Instead, what often determined autonomy was the object's military strength, which either successfully repulsed Chinese aggression or withstood the charge long enough to tire the empire into accepting the limited suzerainty of a tributary relationship. The traditional notion that Chinese identity is based on culture rather than ethnicity facilitated this process of territorial expansion in the Chinese mind. As countries and ethnic groups on China's periphery were invaded and gradually assimilated Chinese culture and customs, it was natural that over time they became incorporated into a new, ever-expanding "Chinese" polity (see "Imperial Chinese State Craft," p. 12).[25]

End of the Imperial Era: China Confronts the West

As mentioned, China during the Ming period began to isolate itself from interaction with the wider world outside its immediate periphery. Following Zheng He's expeditions, increasingly powerful neo-Confucian scholar-bureaucrats advocated a more traditional focus inward on agriculture and cultural purity. Chinese experimentation with expanded foreign contact and influence was terminated. Zhu Di's successors renounced Zheng He's voyages as wasteful and embraced Confucian introspection, leading to a decline in interest in the outside world and in science and technology. Maritime exploration was forbidden, and large ships were ordered destroyed.[26] China ceded its maritime strength to Europe, and by 1503, the Chinese navy was one-tenth the size it had been a century earlier.[27]

For the next few centuries, China remained inward-looking and xenophobic, viewing itself as a model on which other nations should shape themselves, but uninterested in fully engaging in a world that it continued to view as inferior and often threatening. The outside world, the resurgent Confucian bureaucratic elite believed, would taint the superior Chinese culture. Eventually, the outside world, whether inferior or not, would come to China.

Western traders and missionaries had begun to flow into China in the late Ming period of the sixteenth century, despite China's desire for isolation. For the first time, China was confronted by nations with aggressive political, economic, and military cultures that had little patience for Chinese pretensions of centrality or superiority. In the seventeenth century, faced with a military threat from Russia, Imperial China displayed a rather flexible and pragmatic side of its Confucian orthodoxy concerning international affairs. The Qing entered its first international treaty, the Treaty of Nerchinsk, suggesting a relationship between equals. A Qing mission even accepted an opportunity to perform a small kowtow in Moscow, requiring reciprocity of course, in an attempt to prevent Russia from aiding the Mongols who were (as always) challenging the Chinese heartland.[28]

Imperial Chinese Statecraft

A wildcard in any examination of themes in Imperial China's relations with the outside world, particularly one that seeks to detect themes relevant to the present day, is the issue of classical Chinese strategic concepts and statecraft. Despite official Confucian disdain for the military arts, a well-developed strategic culture nonetheless emerged during the earliest days of the Chinese nation that arguably remains embedded within the Chinese popular—and perhaps elite—imagination today. Ancient Chinese statecraft reaching back nearly three thousand years, to the Spring and Autumn (722–481 BCE) and Warring States (403–221 BCE) periods through the dynasties of the first millennium CE, have been immortalized in Chinese classic literature and oral histories that told of the exploits of kings, generals, political strategists, and philosophers as they jockeyed for power and defended the realm during periods of political turmoil in China.[29] This strategic culture included military tactics, but accentuated diplomatic cunning and political maneuver to achieve consolidation of one's power and triumph over one's competitors. As Michael Hunt has written,

> (T)his tradition shows Chinese functioning in an amoral interstate system characterized by constant maneuver and ruthless competition. Its leading figures are not burdened in their decision by hoary tradition; rather, they repeatedly resort to the classic realist calculus, trying to achieve the desired end by the most economical means. Temporary accommodation, alliances made and abandoned, ambush and treachery, the careful cultivation of domestic resources and morale, psychological warfare, and of course raw military power all occupied an important place in the arsenal of the statesmen of these periods of disunity.[30]

The essence of this strategic approach was captured in the so-called "36 Strategems," which used stories from Chinese history—apocryphal or otherwise—to compile a series of pithy sayings instructing the practitioner on how to achieve one's strategic objectives. The sayings are divided into six sections to address situations of advantage and disadvantage, offense and defense. The theme that runs through each strategem, however, is the utility of deception to achieve an objective.[31]

Although developed to address power struggles among warring fiefdoms on the battlefield and in the halls of power, and not necessarily for China's interactions with the wider world, these political and military methods—embedded in popular literature—remain part of China's cultural memory. Mao Zedong, for example, was a notable student and practitioner of classical Chinese statecraft. It is uncertain the degree to which these classical methods are an active part of China's twenty-first century strategic mindset, however the classical tradition remains a potential guide for any modern Chinese (or indeed Western) strategist who seeks to gain strategic advantage over others in a complicated security environment.

Over time, trade restrictions were loosened and tributary relations eased. In the late eighteenth and nineteenth centuries, European powers reached the Middle Kingdom's shores. After 1760, the so-called "Canton system" was established whereby all foreign trade was restricted to a single port at Canton (Guangzhou) to allow the Qing regime to manage the growing European trade.[32] Europe's lust for the China trade, however, led eventually to war to open the China market. Years of isolation and complacent assumption of its own superiority had allowed the empire to weaken both internally and externally, particularly in terms of military power, leaving it unable to meet the new threat from across the sea when it came. China was defeated in the Opium Wars (1839–1842) and compelled to sign the Treaty of Nanjing—the first in a series of "unequal treaties" that completely altered the balance of power between China and the outside world.

The steady loss of domestic sovereignty and territorial integrity at the hands of foreign powers, including Japan, launched China's so-called "century of humiliation" that fundamentally altered China's relationship with its neighbors and dispelled any myth of cultural or any other measure of superiority. The Qing empire fell, ending China's Imperial era. Warlords appeared in its place as well as a Republican era that focused China's energies inward as the country struggled with both a devastating civil war and foreign invasion. Only with the end of World War II in 1945 was the Middle Kingdom able to regain its sovereignty, only to fall into a different type of domestic turmoil—and international orientation—after it "stood up" under Mao's communists in 1949.

Post-1949: The Communist Era

China and the Third World

With the founding of the People's Republic of China (PRC) in 1949, China entered a new phase in its external relations, focusing on relationships not only with the world's superpowers but also specifically with the underdeveloped and developing worlds (which were now defined very specifically according to economic status as opposed to the more culturally-based definition of China's Imperial era). Distinct from its Imperial relationships, these interactions were based on a self-conscious ideology that promoted equality among nations (albeit with some more equal than others) and that pulled China deeply into world affairs.

In a June 1949 speech in commemoration of the founding of the Chinese Communist Party (CCP), Mao Zedong stated that "in order to win victory and consolidate it, we must lean to one side. . . . [A]ll Chinese without exception must lean either to the side of imperialism or to the side of socialism. Sitting on the fence will not do, nor is there a third road." In this same speech, he asserted that China would work to

> . . . unite in a common struggle with those nations of the world which treat us as equals and unite with the peoples of all countries. That is, ally ourselves with the Soviet Union, with the People's Democracies and with the proletariat and the broad masses of the people in all other countries, and form an international united front.[33]

Despite Mao's statement of fealty to communist solidarity and internationalism, however, China's actual policy began to adhere to a "third road," particularly as its relations with both superpowers faltered, that over time had as much to do with practical balance of power politics as ideology. The PRC began to accentuate its relationships with the developing world in a series of stages. In the first stage, China sought to consolidate its position domestically and counter isolation imposed by the United States, which included the establishment of the anti-communist Southeast Asian Treaty Organization (SEATO) on its doorstep immediately following the Korean War. In response, Premier Zhou Enlai—in classical Chinese fashion—worked first to stabilize China's relations with nations along its periphery, signing agreements with the Mongolian People's Republic (Outer Mongolia) and the Democratic People's Republic of Korea (North Korea), and promoting ties with India, Burma, and Vietnamese revolutionaries fighting French colonialism.

Crystallizing China's new diplomatic approach, Premier Zhou Enlai outlined the "Five Principles of Peaceful Coexistence" at a meeting with an Indian delegation on December 31, 1953 as the framework under which its foreign policy would be conducted.[34] The Five Principles—mutual respect for territorial integrity and sovereignty, non-aggression, non-interference in internal affairs, equality and mutual benefit, and peaceful coexistence—all appealed to a developing world that, like China, had felt the brunt of colonialism by Western capitalist powers.

The principles were enshrined the following year at the Bandung Conference in Indonesia. The Bandung Conference, which involved delegates from twenty-nine Asian and African nations, promoted Asia-Africa solidarity and inaugurated China's new role in international diplomacy among the nations of the Third World.[35] Bandung also led to the creation of the "Non-Aligned Movement" in which Third World nations ostensibly refused to side with either the United States or the Soviet Union in what was rapidly becoming a bipolar international environment. China's involvement at Bandung opened the door to greater cooperation with a host of countries around the world that had previously refused to recognize the PRC (in favor of the Republic of China on Taiwan), and offered China an opportunity to assert international leadership.

Maoist Radicalism and the Decline of Third World Solidarity

While certainly a turning point, Bandung and the Five Principles of Peaceful Coexistence were just the first of many theories that defined (and redefined) China's relationship with the developing (and developed) world during the second half of the twentieth century. These theories evolved as China's foreign policy priorities and national interests changed.

In 1963, Mao elaborated on his "Dual Intermediate Zones" theory, which he had first discussed more than a decade earlier.[36] This theory argued that a spacious intermediate zone existed between the United States and the USSR. Countries in the intermediate zone, like China, were independent and should band together to form a united front against imperialist America. The intermediate zone itself was broken down into sections. One part consisted of the developing countries and colonies in Asia, Africa, and Latin America—these countries formed the core of the anti-United States coalition. The other part consisted of Japan, Canada, Oceania, and the capitalist countries of Western Europe—countries that, in the PRC's opinion, were being controlled or negatively influenced by the United States.[37] In this view of interlocking worlds, the developing countries of Asia, Latin America, and Africa were placed at the forefront of a righteous struggle against imperialism. The theory advocated proletarian revolution and economic self-reliance.

If the 1950s were a period in which China formed a united front with moderate Third World nations such as India, Egypt, and Yugoslavia, the 1960s eventually became the "high point of China's sponsorship of Third World radicalism."[38] From 1963 to 1964, Zhou Enlai toured thirteen Asian and African nations to gain support for the Chinese model of socialism. Later, he declared, "The revolutionary movements of the people of the world, particularly in Asia, Africa and Latin America, are surging vigorously forward."[39]

China's anti-colonial rhetoric gave way increasingly to an ideological campaign that included material assistance to communist movements in Third World nations. Beijing sent arms, money, military instructors, and economic aid to help guerrilla movements in developing countries such as Angola, Indonesia, Malaysia, Mozambique, the Philippines, Rhodesia (Zaire), Thailand, and Uganda, and actively fought on behalf of communist insurgents in Vietnam, who received nearly $20 billion in Chinese aid. During the radical height of the Cultural Revolution in the late 1960s, China's leadership ceased to engage with Third World governments in favor of Maoist insurgents inside their countries. Beijing also used its well-honed propaganda machine to spread the communist message in Southeast Asia—through radio broadcasts airing communist propaganda, for example.

Beijing's support for communist insurgent groups predictably soured its relations with many Third World nations, which must have questioned the PRC's actual commitment to the Five Principles of Peaceful Coexistence —particularly concerning non-interference in other nations' internal affairs. Although SEATO had disintegrated, the Southeast Asian nations formed the Association of Southeast Asian Nations (ASEAN) in response to China's provocative support for communist insurgencies in the region.

As before, spreading communism in the Third World was as much a political aim as an ideological one, as China began to compete not only with the United States but also with the Soviet Union following the Sino-Soviet split that became ever wider as the 1960s progressed. This effort to promote Chinese ideology and influence and counter United States and Soviet power was perhaps best articulated by General Lin Biao, a senior party leader at the time. In his famous 1965 essay, "Long Live the Victory of People's War," he predicted that the developing world would usurp power from the superpowers and establish a new world order, with China in the lead.[40]

In the end, however, despite being the largest non-OPEC donor to the developing countries between 1953 and 1985, the PRC had a limited ability to influence Third World developments decisively in their favor during most of the Cold War.[41] Chinese aid to insurgent movements in reality was modest. With the exception of Vietnam, the largest single recipient of Chinese aid was the Pakistani government (ironically in a balance of power struggle against its erstwhile Third World/Bandung ally, India), not African or Latin American rebel groups. The overall Chinese effort in the Third World was disorganized, reflecting the chaos of China's leadership during the period leading up to and during the Cultural Revolution. Rebel groups eventually became disillusioned with their militant Chinese benefactors.

Meanwhile, border wars, with India in 1962 and later with Vietnam in 1979, also revealed the many rifts in Third World solidarity during the Cold War and challenged China's assumption of leadership. China's performance demonstrated a nation as concerned with traditional balance of power considerations and periphery security as extraordinary ideological affiliation with the developing world. China's framing of these interventions as "teaching a lesson" to these border countries also echoed the virtue-based, tributary-style approach to periphery relations that China assumed during its Imperial era—a fact that was not lost on the border states.[42]

Much of China's international attention during this period involved its acrimonious conflict with the Soviet Union, with which it split ideologically and politically in the late 1950s, competed for attention among Third World nations, and fought a hot war along their long common border in 1969. As a result, in the early 1970s China began to rethink its strategic orientation. Beijing looked to reestablish diplomatic ties that had withered or been cut off during the more violent days of the Cultural Revolution. China gradually reentered the community of nations, replacing the Republic of China in the United Nations in 1971 and opening a new era in relations with the United States in 1972.

Nonetheless, even late-Maoist China continued to promote Third World-related theories calling for "anti-imperialist" solidarity and a new international economic order based around the developing world. In a 1974 speech at the United Nations, Deng Xiaoping introduced Mao's "Theory of the Three Worlds"—developed by Zhou Enlai—that divided the world into three zones. The first zone was comprised of the United States and the Soviet Union, whose competition threatened world war. The second zone included the rest of the industrialized world, including Europe, Japan, Canada, Australia, and New Zealand, and the third zone was the non-aligned Third World, led by China. The theory called for the second and third zones to unite against imperialism and an unjust world economic order. While on the surface this statement appeared to be little more than a recapitulation of past theories, Mao's Three World Theory was in fact a significant departure from earlier delineations of the world into socialist and capitalist camps. It abjured ideology and pitted the superpowers against the rest of the world. China made clear its view, however, that the Soviet Union posed a greater threat to the world than the United States, enabling it to continue to pursue closer ties with Washington.[43]

Post-1979: The Decline of Ideology

Beginning in the late 1970s, however, China steadily downplayed the ideological components of both its domestic and foreign policy as it pursued economic liberalization and began to open itself up to foreign trade and investment. China shifted from giving aid to seeking and receiving foreign assistance (although its foreign aid picked up again in the 1980s, particularly to Africa).[44] In 1980, China joined the World Bank and the International Monetary Fund—two of the instruments of the imperialist financial order China once sought to disband. That same year, Japan began its Overseas Development Assistance program in China, both to aid China's reform process and as a form of penance for the damage Tokyo wrought in the 1930s and 1940s. In the 1980s, China became one of the world's largest recipients of World Bank loans ($5.5 billion for 52 projects).[45]

China continued to affirm that its international relations were based on the Five Principles of Peaceful Coexistence, although China's foreign policy was now applied according to practical considerations of economic development, and so on, without regard to a nation's political system or ideology. In 1982, Deng Xiaoping established China's commitment to an "independent foreign policy of peace," a catch-all formulation that sought to reassure the international community of Beijing's new non-aligned and pragmatic orientation (and remains the PRC's slogan for its international policy).

While China accentuated economic development—and its relationship with the United States became a top priority—Beijing continued to seek constructive relationships with Third World countries based on equality and mutual respect, and based its economic relations with the developing world on mutual benefit (what China termed "win-win" results). Because the developing world had little to offer China economically, Beijing's financial aid declined during this period. However, the rhetoric of South-South cooperation continued. Deng did not abandon the notion of people's war, but did change its definition to anything that strengthened the hand of the Third World nations against the United States and other First or Second world powers.[46] China began to provide military aid—including, but not limited to, nuclear technology and missile systems—to Third World nations such as Algeria, Iran, Iraq, Pakistan, Saudi Arabia, and Syria to gain capital, buy influence, and demonstrate its continued fealty and leadership in the developing world. China turned to the developing world in a more concentrated and strategic way following the violent suppression of the Tiananmen Square demonstrations in 1989, when the regime felt the brunt of Western-led condemnation, sanction, and political isolation for its behavior.

During the 1990s, China's continued policy of pragmatism and economic openness paid dividends politically, as it normalized relations and developed closer economic ties with a number of countries throughout the developing (and developed) world. Politically, Beijing continued to pressure developing nations for diplomatic recognition and to adhere in particular to a "One China policy" that meant derecognition of the Republic of China (on the island of Taiwan). Beijing used its growing strength and pragmatic international orientation to isolate Taiwan in the international community, particularly as the PRC began to perceive the island as drifting toward a posture of permanent separation in the late 1990s and early 2000s, in an effort to compel Taiwan's unification with the mainland over time. During the same period, China continued to suspect U.S. motives toward the PRC (and Taiwan), specifically fearing that the United States sought to divide China and contain China's rise. Observing that the end of the Cold War did not lead to a division of power in the world but to heightened American predominance, Beijing began to promote the notion of a multi-polar world to protect its interests and dilute U.S. global power and influence—a posture that resonated well in most of the developing world and led to common cause in international organizations such as the United Nations and the World Trade Organization. With the addition of profound new economic interests in Africa, Latin America, the Middle East, Central Asia,

South Asia, and Southeast Asia, China's strategic attention to the developing world witnessed a renaissance as the twenty-first century dawned.[47]

Conclusion

China's historical sense of superiority over, and responsibility toward, lesser developed neighbors dating back to the tributary period, its experience of victimization and subjugation in the nineteenth and first half of the twentieth centuries, and perhaps modern ambitions for superpower status have led China over much of the last half-century to assert itself as a natural leader of less fortunate nations around the world. Although Chinese leaders no longer take an ideological approach to its relations with the developing world, as during the Cold War, the spirit of Bandung still lives. China continues to promote the Five Principles of Peaceful Coexistence as the foundation of its international diplomacy and strategic rhetoric, whether in the guise of its "New Security Concept"[48] or in its more recent accession to Southeast Asia's Treaty of Amity and Cooperation.

It is clear, however, that China's ideological goals today are different than they were during the Cold War as Beijing, rather than promoting a kind of insurgent revolutionary agenda of ideological hostility to the capitalist world, seeks to attain First World economic and political status with a prominent seat at the table in which global rules are made. It is a goal more reminiscent of, if not precisely akin to, its pre-fifteenth century confident self-image—albeit today in an atmosphere of sovereign equality among states—than its more subversive and insecure Cold War orientation.

China's view of the world and its place in it has evolved greatly over the centuries. While China historically measured itself as a highly developed power largely according to its cultural achievements, economic development and technological prowess have assumed priority today as measurements of China's sense of itself in relation to the world's major powers. This change has had substantial implications on China's current foreign and domestic policy, and places the PRC on a more traditional track in international affairs. For instance, contrary to its Confucian tradition, Beijing has sought in recent years to embrace global trade as an essential priority in achieving national greatness and international prestige, and to develop a military commensurate with its growing economic and political strength. Only too late did Imperial China begin to realize the danger of relying on its sense of cultural superiority and virtue alone to protect itself from the will and depredations of stronger outside powers. It is evident that the China of today has learned from that historical lesson.

China's traditional perception of its "strategic periphery" has also grown due to modern realities. While once this periphery encompassed Tibet and the nomadic areas of the northern and northwestern steppe, today such areas are considered proper Chinese territory, requiring a new and expanded definition of China's overland strategic requirements. Today China must also look south and east toward the security of its maritime coastline. This includes the island of Taiwan, whose deep emotional resonance to China transcends security, as well as the East China Sea and South China Sea—each of which involves elements of both military and economic security.[49]

One might argue that in today's globalized world, China's urgent domestic needs extend its strategic periphery even farther to those areas that may offer the markets, investment destinations, and natural resources that are required to keep China developing economically so Beijing may maintain domestic stability. Contrary to its Imperial days, China's days of self-sufficiency are past, and it requires urgent interaction with an array of nations—many of which are in the developing world and include unsavory regimes. How will China balance its own strategic interests with the interests and values of the international community as a whole?

As described in this chapter, China has had many different historical experiences with the outside world that may inform its future approach to the developing world. Ancient Chinese statecraft, passed down from the earliest days of China's Imperial history, reflects a tradition of deception and intrigue and a highly developed sense of political strategy that enabled it to gain advantage over rivals. The Tang and Song periods offer another tradition in which relations between China and the outside world, while based on China's sense of its own cultural, philosophical, and technological superiority, were relatively open to the neighboring states with which it traded and engaged productively to mutual benefit.

Finally, there is the most well-known historic tradition of tributary relations with outside powers, in which China's strong sense of its own virtue and (cultural) superiority dominated its interactions with the outside world (and from which outside powers gained materially, as well). That tributary situation was established originally as a holding action for China during periods of weakness and vulnerability, when its borders were subject to threats from China's strategic periphery and China's leadership was eager to focus on domestic instability in the heartland. This posture would give way periodically, however, to a more insidious and sometimes aggressive expansionism when China regained its strength and felt more secure at home. One might detect echoes of each of these three traditions in China's current engagement with the international community in general, and the developing world in particular.

So how might China's past inform its future? Will China become an insidiously aggressive and acquisitive major power, or merely a benign nation with pretensions of superiority? Is China's fundamental international orientation one of dominance or defense? How does the tradition of "rule by virtue" connect with China's modern attention to principle in framing its conduct of international affairs, and how does "teaching lessons" connect to principle in its conduct of military campaigns? How might new self-images, related for instance to its understanding of China's history, territory, or international stature, inform its future perspective on regional and global affairs—and perhaps challenge the interests of other states? And when the modern Chinese imperium becomes strong militarily and stabilizes its domestic situation, where might its energies then be directed?

No one can answer these questions with certainty given the transitional nature of China's rise and the many uncertainties inherent in considering China's future. China clearly does not have the same easy self-confidence and self-regard it had during its Imperial days (although its confidence and nationalism is indeed rising along with its per capita income). Even if it regained such self-regard, Beijing faces new constraints on its ability to dominate its neighbors or assert its cultural superiority in the same way it did when China's emperors believed they ruled "All Under Heaven." Among others, perhaps the most compelling difference today is the presence of another proud, self-reverential, often arrogant nation with its own sense of exceptionalism that considers itself the world's Middle Kingdom in all but name. Managing its relationship with, and the influence of, the United States around the world, particularly along China's border, will be Beijing's most critical challenge. Likewise, how the United States handles the emergence of history's Middle Kingdom, including the complexities of China's growing interactions in the developing world, will go far to determining the course of international security in the coming years.

Notes

1. For an extensive account of the seven expeditions by Zheng He, see Roderich Ptak, *China and the Asian Seas: Trade, Travel, and Visions of the Other (1400–1750)* (Aldershot, Brookfield: Ashgate, 1998), 97–107.

2. Bruce Swanson, *Eighth Voyage of the Dragon: A History of China's Quest for Seapower* (Annapolis, MD: Naval Institute Press, 1982); John R. Dewenter, "China Afloat," *Foreign Affairs* 50, no. 4 (July 1972): 738–751; and Bernard Cole, "Waterways and Strategy: China's Priorities," *China Brief* 5, no. 4 (February 2005): 1–3.

3. Jonathan Spence, *The Search for Modern China* (New York: W.W. Norton & Company, 1990), 119.

4. Mark Mancall, *China at the Center: 300 Years of Foreign Policy* (New York: Free Press, 1984), 7.

5. John King Fairbank and Merle Goldman, *China: A New History* (Cambridge, MA: The Belknap Press of Harvard University Press, 1998), 108–9.

6. Examples include the Xiongnu during the Han Dynasty, the Tibetan kingdom during the Tang, and the Ly Dynasty of Vietnam during the late Tang/early Song period. Swaine and Tellis, *Interpreting China's Grand Strategy: Past, Present, and Future*, 54.

7. It is revealing about the Confucian mindset toward the outside world to note that merchants were placed on the lowest rung of the Confucian social hierarchy. Traders were considered too parasitic and nonintellectual for the Confucian scholar. Qing-era Chinese who traveled to or traded with the outside world, in fact, were viewed as having abandoned their country. Nonetheless, Chinese merchants were leaders in China's historical outreach to the world, in the Asian region and beyond. See Spence, *The Search for Modern China*, 119.

8. The Qing emperor Qian Long's famous letter to George III in 1792 responding to British emissary Lord George Macartney's mission to establish closer trade relations between the two countries encapsulated with rare efficiency the historical attitude of Chinese emperors toward such entreaties:

> We have never valued ingenious articles, nor do we have the slightest need of your country's manufactures. Therefore, O king, as regards your request to send someone to remain at the capital, while it is not in harmony with regulations of the Celestial Empire we also feel very much that it is of no advantage to your country.

Needless to say, Macartney left without achieving trade concessions. Spence, *The Search for Modern China*, 122-23; Mark Mancall, "The Ch'ing Tribute System: An Interpretive Essay," in *The Chinese World Order: Traditional China's Foreign Relations*, ed. John King Fairbank (Cambridge, MA: Harvard University Press, 1968), 89.

9. Swaine and Tellis, *Interpreting China's Grand Strategy: Past, Present, and Future*, 22–3.

10. This "pitting barbarian against barbarian" strategy was common throughout Chinese history and is a key element of classical Chinese statecraft, as discussed below. During the Han period, the strategy was manifested in its unsuccessful attempts to enlist peoples in Central Asia as allies to combat the Xiongnu threat along its border, while the Ming's similar "divide and rule" strategies sought to pit potential enemies against one another to make them less threatening to, and perhaps more susceptible to, absorption by the Chinese state. Indeed, many will note echoes of this strategy in communist China's relations with the United States and Soviet Union during the later years of the Cold War. John King Fairbank and Edwin O. Reischauer, *China: Tradition and Transformation* (Boston: Houghton Mifflin, 1973), 63; Geoff Wade, "Ming China and Southeast Asia in the 15th Century: A Reappraisal," *Working Papers Series 28* (Singapore: National University of Singapore Asia Research Institute, July 2004): 39, 24.

11. Fairbank and Goldman, *China: A New History*, 61.

12. The Chinese themselves never actually used the terms "tribute" or "tributary system" to describe their interactions with outside powers. In fact, there is no such Chinese word or phrase. The two terms are Western appellations. To the Chinese, such interactions were characterized as that between "civilization" and "barbarism," with a corresponding lexicon that reflected this state of relations. Gungwu Wang, "Early Ming Relations with Southeast Asia," in *The Chinese World Order: Traditional China's Foreign Relations*, ed. John King Fairbank (Cambridge, MA: Harvard University Press, 1968), 41–43; Mancall, "The Ch'ing Tribute System: An Interpretive Essay," 63.

13. By the nineteenth century, under the tribute system, Korea paid tribute once a year; the Ryukyu Kingdom (today's Okinawan islands) once every two years; Annam (Vietnam) once every three years; Siam (Thailand) once every four years; the Philippines once every five years; and Burma and Laos once every 10 years. John K. Fairbank, "A Preliminary Framework," in *The Chinese World Order: Traditional China's Foreign Relations*, ed. John K. Fairbank (Cambridge, MA: Harvard University Press, 1968), 11. It is noteworthy that feudal Japan alone rejected tributary status, although by the early Qing Dynasty its leaders had chosen to withdraw into seclusion and trade with China only indirectly. Nonetheless, Japan did model its culture after China, again offering the Middle Kingdom reaffirmation of its superiority over those in its orbit. John Miller, "The Roots and Implications of East Asian Regionalism," *Occasional Paper Series*, Asia-Pacific Center for Security Studies (September 2004): 11.

14. China periodically demonstrated its seriousness in upholding the reciprocal nature of tributary responsibility. For instance, Chinese forces helped Korea expel invading Japanese forces in 1592, fulfilling its side of the bargain in protecting a close tributary state (although one might argue that more practical calculations concerning the security of China's northeast frontier were more decisive, as they were in 1950). In 1788, the Chinese went to the aid of the ruling Le Dynasty in Vietnam, citing their right and obligation to do so under the tributary system, but later switched sides to the Nguyen family when the latter proved more committed to its tributary obligations. Spence, *The Search for Modern China*, 119.

15. It should be noted that this examination excludes consideration of the Yuan (Mongol) Dynasty, whose alien incorporation of China into an intercontinental empire, and singular culture of aggression, is considered an anomaly in Chinese history.

16. Fairbank and Reischauer, *China: Tradition and Transformation*, 63.

17. Ibid., 96.

18. Swaine and Tellis, *Interpreting China's Grand Strategy: Past, Present, and Future,* 52, footnote 46.

19. Ibid., footnote 47.

20. Among the far-flung places Admiral Zheng reached were Java and Sumatra (modern Indonesia), Ceylon (Sri Lanka), Siam (Thailand), Vietnam, Cambodia, India, Bangladesh, Yemen, Arabia, Somalia, and Madagascar.

21. Fairbank and Reischauer, *China: Tradition and Transformation*, 198; Philip Bowring, "China's Growing Might and the Spirit of Zheng He," *International Herald Tribune* (August 2, 2005).

22. *Tai-zu shi-lu, juan*, as cited in Wade, "Ming China and Southeast Asia in the 15th Century: A Reappraisal," 39 (1b).

23. Wade, "Ming China and Southeast Asia in the 15th Century: A Reappraisal," 22.

24. Evelyn S. Rawski, "Reenvisioning the Qing: The Significance of the Qing Period in Chinese History," *Journal of Asian Studies,* vol. 55, no. 4 (November 1996): 829–50; Swaine and Tellis, *Interpreting China's Grand Strategy: Past, Present, and Future*, 60–61.

25. Today's Outer Mongolia makes an interesting exception to this rule, although one might argue that Mongolia's independence was an accident of history due more to the legacy of Soviet communism and Cold War dynamics than to any inherent difference in Mongolia's historical status with China. The leaders of Mongolia today surely recognize this situation as they deal delicately with growing Chinese influence in their economy and society.

26. Ptak, *China and the Asian Seas: Trade, Travel, and Visions of the Others (1400–1750)*, 105.

27. Spence, *The Search for Modern China*, 119.

28. Fairbank and Reischauer, *China: Tradition and Transformation*, 253.

29. These volumes include Luo Guanzhong's "The Romance of the Three Kingdoms," Sima Qian's "Records of the Grand Historian of China," and other Chinese dynastic histories. Sun Zi's classic "The Art of War" remains a standard exposition of classic Chinese battlefield tactics and military strategy. For an excellent and entertaining exposition of the various competing political philosophies and military strategies developed during this period of ancient China, see in particular Dennis Bloodworth and Ching Ping Bloodworth, *The Chinese Machiavelli: 3000 Years of Chinese Statecraft* (Somerset, NJ: Transaction Publishers, 2004). For an authoritative and landmark study of China's strategic culture, see Alastair Iain Johnston, *Cultural Realism: Strategic Culture and Grand Strategy in Chinese History* (Princeton, New Jersey: Princeton University Press, 1998).

30. Michael H. Hunt, "Chinese Foreign Relations in Historical Perspective," in *China's Foreign Relations in the 1980s*, ed. Harry Harding (New Haven, CT: Yale University Press, 1984), 7–8.

31. Examples of the stratagems include: "Befriend a distant state while attacking a neighbor"; "Wait at ease for the fatigued enemy"; "Sacrifice the plum for the peach"; "Make a feint to the east while attacking in the west"; and "Inflict injury on oneself to win the enemy's trust." See Koh Kok Kiang and Liu Yi, trans., *The Thirty-Six Stratagems: Secret Art of War* (Singapore: AsiaPac Books, 1992).

32. For a comprehensive summary of the process of institutionalizing the Canton system, see Earl H. Pritchard, *The Crucial Years of Early Anglo-Chinese Relations, 1750–1800* (New York: Octagon Books, 1936), 128–141.

33. Mao Zedong, "On the People's Democratic Dictatorship, June 30, 1949," in *Mao Zedong Xuanji* (Selected works of Mao Zedong) (Beijing: People's Press, 1991), vol. IV, 1477.

34. Zhou Enlai, *Zhou Enlai Waijiao Wenxuan* (Selected works of Zhou Enlai on diplomacy) (Beijing: Central Documents Press, 1990), 63.

35. Coined by economist Alfred Sauvy in a 1952 article in the French magazine *L'Observateur* as a deliberate reference to the Third Estate of the French Revolution, the term "Third World" was originally used to describe those nations that were nonaligned in the Cold War, "those less-industrialized countries opposed to political and economic domination by the superpowers and the developed world." *L'Observateur*, no. 118 (August 14, 1952). See also Anthony I. Akubue, "Gender Disparity in Third World Technological, Social and Economic Development," *Journal of Technology Studies,* vol. 27, no. 2 (2001): 64. The term came to prominence following the Bandung Conference when developing nations explored ways to counter a world dominated by "American imperialism" and, to a lesser extent, the Soviet Union.

36. The theory was first presented during a talk between Mao and American journalist Anna Louise Strong in 1946. See Mao Zedong, *Mao Zedong Xuanji*, vol. IV, 1194.

37. Mao Zedong, "There Are Two Intermediate Zones," *Mao Zedong Waijiao Wenxuan* (Selected works of Mao Zedong on diplomacy) (Beijing: Central Documents Press, 1994), 506–509.

38. Donald W. Klein, "China and the Second World," in *China and the World: New Directions in Chinese Foreign Relations*, ed. Samuel S. Kim (Boulder, CO: Westview Press, 1989), 130.

39. Premier Zhou Enlai's Speech at the National Day Reception, September 30, 1966.

40. Lin Biao, "Long Live the Victory of People's War!" *Peking Review* (September 3, 1965). During this period, China also supported Indonesia's call for a new United Nations for developing nations. Harry Harding, "China's Changing Roles in the Contemporary World," in *China's Foreign Relations in the 1980s*, 187.

41. Organisation of Economic Co-operation and Development, *The Aid Programme of China* (Paris: OECD, March 1987), 5.

42. China's intervention in the Korean War in 1950 might also be viewed as an echo of China's long-time concern about the vulnerability of its periphery, albeit without the formal tributary or "virtue-based" components. For an excellent discussion of how Chinese communists have framed their military campaigns in terms of "teaching a lesson," see Allen S. Whiting, "China's Use of Force, 1950–96, and Taiwan," *International Security*, vol. 26, no. 2 (Fall 2001), 103–131.

43. For instance, "Speech by Deng Xiao-ping, Chairman of Delegation of People's Republic of China at Special Session of United Nations General Assembly," *Peking Review* (September 1974), I–II.

44. Organisation of Economic Co-operation and Development, *The Aid Programme of China*, 5.

45. Klein, "China and the Second World," 163.

46. Thomas W. Robinson, "Chinese Foreign Policy from the 1940s to the 1990s," in *Chinese Foreign Policy: Theory and Practice,"* ed. Thomas W. Robinson and David Shambaugh (Oxford: Clarendon Press, 1994), 575.

47. Council on Foreign Relations, *More than Humanitarianism: A Strategic U.S. Approach Toward Africa*, Report of an Independent Task Force (January 2006): 42.

48. Beijing developed its "New Security Concept" in the mid-1990s to challenge the notion that international security, particularly in East Asia, requires the underlying guarantee of U.S.-based alliances. The concept stresses development of informal strategic partnerships between nations, adherence to the Five Principles of Peaceful Coexistence, and multilateralism, which itself promotes a more equal role for the developing nations in setting the rules of international affairs. See *Zhongguo Guanyu Xin Anquanguan de Lichang Wenjian* (Document on China's position on the new security concept), submitted by the Chinese delegation to the ASEAN Regional Forum on July 31, 2002, *People's Daily* (August 2, 2002): 3.

49. "China's Worries At Sea," *Global Times* (reprinted in the *People's Liberation Daily*) (January 2, 2004), available at http://www.uscc.gov/researchpapers/2004/chinaworriesatsea.htm.

II

Regional Profiles

2

China's Post-Cold War Strategy in Africa

Examining Beijing's Methods and Objectives

Joshua Eisenman

Over the last decade, the People's Republic of China (PRC) has constructed a foreign policy toward Africa designed to secure natural resources, consumer markets, and its position as leader of the developing world. To realize these goals, Beijing has employed approaches that address the African nations' economic, diplomatic, and security needs, while ensuring China's continued ability to influence the political and commercial landscape of this resource-rich continent. China's increasingly proactive foreign policymakers have taken advantage of a void left by an indifferent Russia, a preoccupied United States, and a divided Europe, to create fresh opportunities and pursue new bilateral and multilateral dialogues.

For the United States, China's strategy in African affairs has unique implications. Washington—preoccupied by and economically committed to a costly war on terror and the reconstruction of Iraq and Afghanistan—has allowed Africa to remain low on its foreign policy priority list. As a result, until recently Beijing has been able to pursue its strategy in Africa without drawing much attention and has expanded the depth and breadth of its political, economic, diplomatic, and military relationships with African leaders in a manner that may have adverse implications for U.S. interests in the region.

This chapter will begin with a brief historical overview of Sino-African relations. It will explore the development of Beijing's strategy, with particular attention to China's contemporary objectives and methods, while highlighting key bilateral relationships and the implications of China's approach for African nations and the United States.

Background

In the 1950s and 1960s, Beijing's primary motivation in Africa was the affirmation of its own brand of communism and support for revolutionary

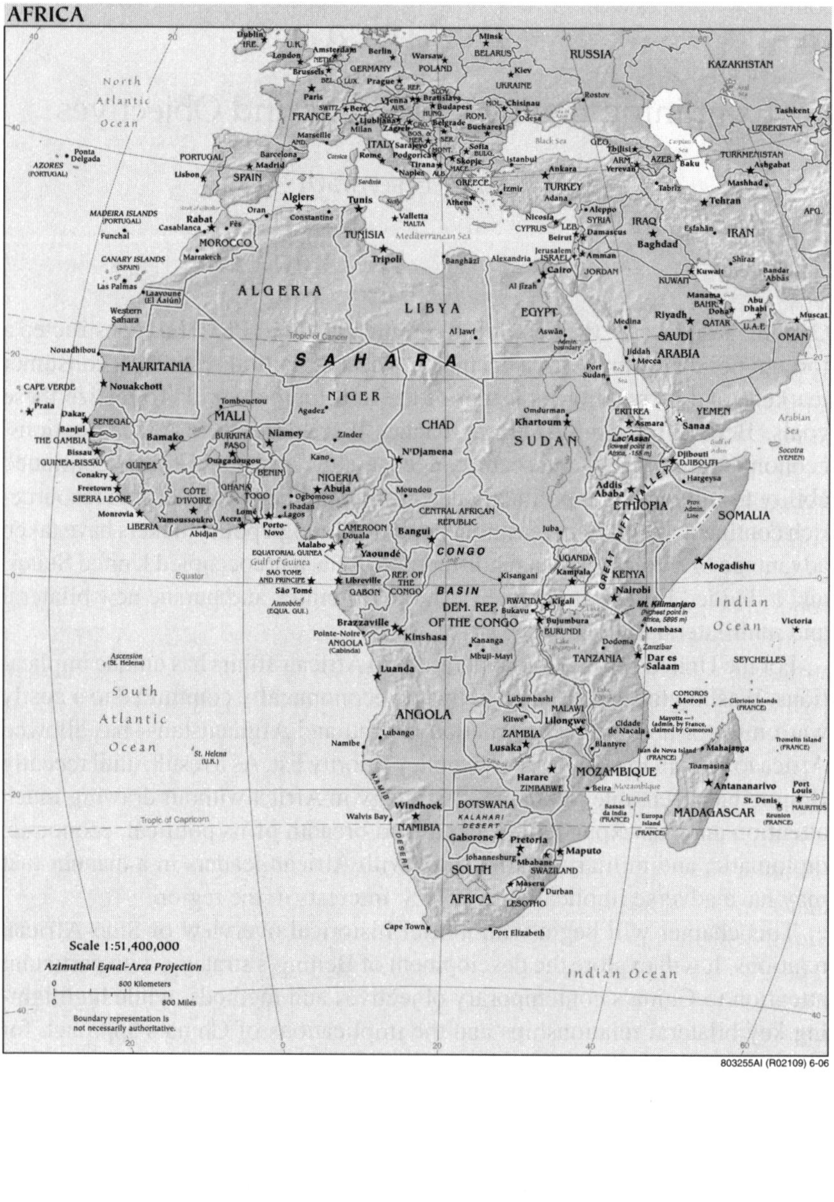
AFRICA
North Atlantic Ocean
South Atlantic Ocean
Indian Ocean
Mediterranean Sea
Gulf of Guinea
Equator
Tropic of Cancer
Tropic of Capricorn
SAHARA
ALGERIA
LIBYA
EGYPT
MOROCCO
TUNISIA
Western Sahara
MAURITANIA
MALI
NIGER
CHAD
SUDAN
ERITREA
DJIBOUTI
ETHIOPIA
SOMALIA
SENEGAL
THE GAMBIA
GUINEA-BISSAU
GUINEA
SIERRA LEONE
LIBERIA
CÔTE D'IVOIRE
BURKINA FASO
GHANA
TOGO
BENIN
NIGERIA
CAMEROON
CENTRAL AFRICAN REPUBLIC
EQUATORIAL GUINEA
SAO TOME AND PRINCIPE
GABON
REP. OF THE CONGO
DEM. REP. OF THE CONGO
CONGO BASIN
UGANDA
KENYA
RWANDA
BURUNDI
TANZANIA
ANGOLA
ZAMBIA
MALAWI
MOZAMBIQUE
ZIMBABWE
NAMIBIA
BOTSWANA
KALAHARI DESERT
NAMIB DESERT
SOUTH AFRICA
SWAZILAND
LESOTHO
MADAGASCAR
COMOROS
SEYCHELLES
MAURITIUS
CAPE VERDE
SPAIN
PORTUGAL
FRANCE
ITALY
GREECE
TURKEY
SAUDI ARABIA
YEMEN
OMAN
IRAQ
IRAN
RUSSIA
KAZAKHSTAN
Scale 1:51,400,000
Azimuthal Equal-Area Projection
800 Kilometers
800 Miles
Boundary representation is not necessarily authoritative.
803255AI (R02109) 6-06

movements. In the 1970s, following the most tumultuous periods of Mao Zedong's Cultural Revolution and the deepening of the Sino-Soviet split, an increasingly pragmatic leadership looked to secure China's geopolitical interests and its borders by keeping Soviet resources bogged down in distant conflicts. Today, to support China's growing economy, Beijing is developing extensive commercial and diplomatic ties with Africa.

Mao's Approach

During Mao Zedong's rule, the Chinese Communist Party (CCP) brought African policy to the Chinese people via the state-run press. To reinforce domestic support and publicize their conviction, the CCP used newspapers to broadcast China's role, promoting what Mao called "righteous struggle" in Africa. This meant supporting Mao-style revolutionary mass movements in Africa as an extension of China's own unfinished revolution.[1] By citing Africa as proof of the widespread appeal of Mao Zedong Thought, China aimed to reinforce its ideology's broad appeal and establish its position as the guardian of global proletarian revolutionary orthodoxy.

Beijing aided any African revolutionary force fighting a guerrilla war—a hallmark of the Chinese style—by hastening "the development of [African] political opposition groups and guiding them towards conceptions of action closely akin to her own."[2] China laid claim to the moral high ground by condemning vestiges of Western colonialism. Speeches, editorials, and publications stressed the role of Mao Zedong Thought and the scope of armed struggles.[3] By asserting that conflicts in Algeria, Cameroon, the Congo, Uganda, and elsewhere were proletarian revolutions, China showcased its influence in the region. Calls for armed struggle did not cost China much, so if an indigenous group chose rebellion, Beijing would support it with zealous rhetoric and modest arms shipments.[4]

China portrayed itself as shepherd of a flock of African nations moving toward a "new democratic revolution." Premier Zhou Enlai nurtured the idea that Africa was engulfed in a wave of revolutionary zeal. At the Moscow Summit of Communist Parties in November 1960, China's state-run press reported that African revolutionaries were "studying Mao's works and using Chinese guerilla methods."[5] Indeed, many African leaders were steeped in Maoist revolutionary thought and liberation ideologies.[6]

The Birth of Chinese Pragmatism

In 1969, Beijing's greatest fears seemed on the verge of realization. Roughly 400,000 Soviet troops equipped with battlefield nuclear weapons appeared on the Sino-Soviet border. Military intelligence officers in Washington predicted a massive Soviet invasion.[7] The Soviets never launched a full offensive, but the threat prompted Beijing to devise a strategy to cope with the Soviet menace. This new approach was based on pragmatism rather than ideology, and signaled the start of a strategy in Africa and elsewhere based on Beijing's national security interests—a theme that continues to this day.[8]

This shift from dogmatism to pragmatism was catalyzed by widespread cynicism as the Cultural Revolution's worst days subsided. The largely disillusioned Chinese leadership, weary of ideological fervor and fearful of Soviet aggression, turned to geopolitical realism rooted in self-preservation to propel policymaking. In this way, the CCP's ideology of continuous proletarian revolution was gradually altered in favor of a theory of anti-hegemony.

Although Beijing publicly called for a dual-adversary approach directed against both the United States and the USSR, in practice the Soviet Union was singled out as the greater threat.[9] Beijing began to support only those revolutionary movements that fought against "imperialist forces"—a term synonymous with groups supported by Moscow and, to a lesser extent, Washington. China's willingness to place its geopolitical objectives before ideological consistency grew apace with the Soviet threat. The result was a strategy designed to preoccupy Soviet forces in far-off conflicts—particularly in Africa. This method was not new, but its value grew exponentially with the Soviet threat. Richard Lowenthal explains this approach:

> From the Chinese point of view, to multiply the scenes of local fighting far from China's shores means not to multiply the risks but to multiply the diversions. The Chinese have a positive interest in local conflicts further away, and among other things in local conflicts in Africa.[10]

Angola was one country in which Beijing aimed to undermine Moscow's influence in Africa. Beijing, which had first supported the Popular Movement for the Liberation of Angola (MPLA), was swayed by Zaire's (now the Democratic Republic of Congo) support for Holden Roberto's National Front for the Liberation of Angola (FNLA).[11] In December 1973, Deng Xiaoping met with FNLA representatives, and six months later "hundreds of tons of arms were flowing to Roberto's army via Zaire and a 112 man contingent to train guerrillas had arrived. By 1975, the FNLA had achieved military superiority over the MPLA."[12] The decision to aid the FNLA was not only an important change in China's Angola policy, but it also reflected a willingness to abandon an openly communist group—the MPLA—simply to keep Moscow's forces dispersed. Chinese arms, funds, and training forced the Soviet Union to act decisively to secure its interests in Angola. Soviet aid to the MPLA was $27 million from 1960 to 1974, but totaled $100 million in 1975 alone.[13]

After Mao's death in 1976 and the fall of the Gang of Four, the CCP's pragmatists led by Deng Xiaoping gradually emerged. Although Mao had already been enshrined as "the great leader of the international proletariat and the oppressed nations and the oppressed people,"[14] his followers now suffered from a crisis of conscience. The belief structure to which they had subscribed failed to fulfill its promise as a panacea for all of society's ills. Free from the need to describe all rebel movements as Maoist, and desperately poor, China began to promote African self-reliance. Beijing became willing "to grant ideological autonomy, and when African countries seemed to embark on a policy closely akin to Chinese thinking, Peking refrained from claiming that the Africans were following a Maoist path."[15] This strategy allowed for an increase in state-to-state relations and put most leftist radicals on notice that they could not expect much support from Beijing. China's attempt to balance a variety of strategic objectives resulted in a seemingly contradictory dual track policy: official diplomatic relations with some African states improved, while China continued to support militant liberation struggles in several sub-Saharan nations.

In the 1980s, changes in China's domestic landscape diverted Beijing's attention from Africa. By the time the pragmatists had completely wrestled the reigns of power from the Maoists in 1980–81, rural poverty, economic stagnation, and widespread public disenchantment required a reorientation of priorities. Led by Deng Xiaoping and Zhao Ziyang, China's leadership was focused on gradually introducing market forces into the economy. With development now placed at the fore, Chinese people were told to get rich, leading many to turn to trade with the West to make their fortunes. This economic dynamic, a receding Soviet threat, and the waning role of revolutionary ideology diverted Beijing's attention from Africa. It was not until the nation's need for raw materials and support on critical diplomatic issues from Taiwan to human rights that Beijing turned back to Africa in the post-Cold War years.

China's Present-day Africa Strategy

Today, Mao's Little Red Book has been replaced by a balance sheet. Africa is now a component in China's larger strategy to cultivate political support, bolster its claims to Taiwan, acquire energy and natural resources, and secure its commercial interests. Beijing also sees the African nations as valuable backers of its claims to lead the developing world and in the struggle against American "hegemony."

China's leaders and state-owned firms have actively courted African political, military, and commercial leaders. Without a free press or much public debate over foreign policy, China's leaders have a free hand to pursue their goals in Africa. China's economy and growing industries demand increasing amounts of energy and raw materials; its exporters want new markets; its diplomats require support in international organizations and new allies to advance PRC interests and, when necessary, counter the United States. The two sections below describe these strategic objectives as well as the methods Beijing employs to achieve them.

Objectives

Political Support

Increasing its influence within international institutions is a primary objective of Beijing's national security strategy, and thus, one of Beijing's primary objectives in Africa. One way China has done this is by using rhetoric intended to fuse the interests of developing nations with its own and then collaborating to pursue those mutual interests in international institutions.[16] In Africa, this rhetoric is based on decades of Sino-African cooperation, which provide an important basis for CCP propagandists' claims. It is by framing the developing world's critique of Western-dominated geopolitics that China seeks to secure its role as leader of an alliance of developing nations.[17] In April 2005, at a speech before the Asia-Africa Business Forum, Chinese President Hu Jintao made this claim:

> As a developing country itself, China takes the strengthening of its friendship and cooperation with developing countries as the cornerstone of its foreign policy and always attaches great importance to enhancing its economic and trade cooperation with Asian and African countries.[18]

By calling its ties to developing countries a "cornerstone" of Beijing's foreign policy, President Hu was unequivocal about Africa's important role. This support is harnessed to achieve China's objectives, whether they are countering U.S. influence on the international stage or deflecting criticism of Beijing's human rights abuses.

Beijing regularly refers to U.S. foreign policy as hegemonic.[19] In June 2005, for example, the *People's Daily* claimed, "U.S. hegemony and unilateralism have experienced malignant swell."[20] This claim is also made in African states. In December 2003, during a ceremony to open the second ministerial meeting of the China-Africa Cooperation in Addis Abada, Ethiopia, Forum Premier Wen Jiabao said "The people of the world share the aspiration for peace, stability and development. But hegemonism is raising its ugly head."[21]

In this way, Beijing is effective in drawing a distinction between itself and the United States. This serves Beijing's interests, but does so at the expense of U.S. efforts on the continent.

China has formalized its political influence through the China-Africa Cooperation Forum. Established in 2000, the Forum allows Beijing to institutionalize its diplomatic overtures, technical training, debt relief, loans, grants, and infrastructure projects. Politically, the Forum allows Beijing and African nations to speak with one voice. For instance, at the Forum's first session in Beijing, eighty African leaders from forty-four nations joined Jiang Zemin's call for a new world order and redress for the unfairness of globalization.[22] And at the Forum's second ministerial meeting held in the Ethiopian capital of Addis Ababa in 2003, the members agreed on the Addis Ababa Action Plan, which reinforced China's presence on the continent.

In November 2006, China hosted the third ministerial meeting in Beijing and raised the level of dialogue with its first ever China-Africa Summit. Forty-eight African leaders attended the summit, which concluded with Hu Jintao's pledge to distribute $5 billion in loans to Africa over three years, and create a $5 billion China-Africa development fund to encourage Chinese companies to invest in Africa.[23]

The United Nations is another forum in which China enjoys the support of African nations. The best example of this was the United Nations Commission on Human Rights (UNCHR), which was replaced in May 2006 with the UN Human Rights Council. African support played a key role in Beijing's successful, eleven-year campaign to avoid UNCHR censure for its human rights record.

Beijing began focusing on the UNCHR in March 1995, when it lost a motion to quash a vote on a resolution supported by Western nations to condemn China's human rights practices. Although China and its allies won, the vote on the resolution was close, leading Beijing to begin a campaign to influence the African voting members of the Commission.[24] As a part of this effort, Vice Premier Li Lanqing visited six African countries, including UNCHR members Mali, Guinea, Gabon, Cameroon, and Côte d'Ivoire. Not surprisingly, after China won the April 1996 UNCHR vote, all fifteen African members held high-level talks with China and signed at least twenty-three agreements and protocols on cooperation.[25]

In April 2004, China again put forth a motion designed to avoid censure for human rights abuses, and it also passed with the support of every African nation on the UNCHR. Of the twenty-seven countries that supported China's motion, over half were African, including nations with questionable human rights records such as Sudan, Eritrea, and Zimbabwe—each of which had its membership supported by Beijing.

For Beijing, the UN Human Rights Council promises an equally hospitable environment. While Zimbabwe and Sudan have been removed, thirteen African nations—many of them with extensive commercial and political relations with Beijing—remain on the Council. This ensures that China will continue to exert influence and that the Council will not take actions contrary to Beijing interests. It will also allow African nations to continue to leverage their votes as political capital in their bilateral negotiations with Beijing.

Taiwan

China enjoys the vast majority of African nations' support in its efforts to isolate Taiwan from the international community. The PRC's rivalry with the Republic of China (ROC) on Taiwan stems from the Chinese Civil War in 1949, when Chiang Kai-shek's Nationalists lost to Mao Zedong's Communists and fled to the island. Between 1949 and 1991, both Taipei and Beijing claimed to be the sole legal authority of all China, and vied for diplomatic recognition to support their claims.[26] Africa remains a key battleground for Beijing and Taipei's so-called "dollar diplomacy," as they compete to establish ties with as many nations as possible.

At nearly every meeting with their Chinese counterparts, African leaders are expected to publicly reaffirm their support for Beijing's "one-China principle." Many African nations see affirmation of China's position as a necessary part of their diplomacy with China, one that costs them nothing but can pay dividends. Some African nations have even taken the initiative. For example, in March 2005, Ethiopia's parliament approved a resolution in support of Beijing's Anti-Secession Law.[27]

Back in the early 1990s, Taipei's financial incentives and able diplomacy helped it to establish formal relations with several African countries at Beijing's expense. Senegal, a long time ally of Beijing and one of West Africa's most important states, switched sides. For nearly thirty years, the PRC had cooperated with Senegal, helping to build hospitals and even a huge national stadium. Taipei also used its financial resources to woo Gambia and Niger, providing the former roughly $35 million in assistance and helping the latter pay civil service salaries.[28]

Unfortunately for Taipei, the fruits of dollar diplomacy have been short-lived. In 1996, Niger broke with Taipei, and in late 2005 Beijing lured Senegal back to its camp. In August 2006, Chad also severed ties after Beijing agreed to stop supporting the country's rebels, according to Taiwan sources.[29] But Taipei's biggest loss came in 1998, when South Africa, sub-Saharan Africa's largest and wealthiest country, exchanged ambassadors with Beijing.

Today, only five African countries maintain formal diplomatic relations with Taiwan, although they represent about one-quarter of all Taiwan's formal diplomatic partners. They include Burkina Faso, Gambia, Malawi, Sao Tome and Principe, and Swaziland. Taiwan continues to supply these nations with funds, but also hopes that increased cooperation in developing communications technology, democratization, and good governance will entice African nations.[30] In addition to bilateral ties, Taiwan also coordinates its efforts in Africa through the Africa-Taiwan Economic Forum and Taiwan International Health Action (TaiwanIHA). These organizations provide Taipei's African supporters with health services and supplies as well as "Taiwan's well-developed technological and international marketing expertise."[31] Unfortunately for Taipei, these tactics are increasingly undermined by China's rapid economic development and political power, leading many African leaders to realize that their historic ties to the ROC are no longer as valuable as relations with the mainland.[32]

For the PRC, gaining international support for the "one-China principle" remains a top priority and Beijing's only prerequisite for official recognition. In October 2003, for instance, after nearly twenty years as an ally of Taiwan under dictator Charles Taylor, Liberia reestablished relations with the PRC. Liberia's new leaders—whose priority was reconstruction following years of civil war—were swayed by Beijing's combination of economic and diplomatic clout. A spokesman for the Liberian president said switching recognition to Beijing would help with Liberia's reconstruction efforts.[33] He was referring to the aid package Beijing leveraged to sway the embattled nation. China agreed to support a United Nations resolution to budget $250 million for 15,000 peacekeepers to stabilize Liberia.[34] It also offered duty-free status for all Liberian imports, $5.6 million in interest-free loans and grants for the Liberian military, and assistance and training in energy development, infrastructure, agriculture, and manufacturing.[35] Beijing had the power to veto the UN resolution, had Liberia's government not acquiesced, providing another powerful incentive for Monrovia to recognize Beijing.

This strategy was again on display in August 2006 when Chad, desperately seeking support on the UN Security Council, and in its dealings with the World Bank, also switched recognition to Beijing. Chad has seen 200,000 refugees arrive on its soil and faces internal strife stemming from the conflict in the Darfur region of western Sudan. This has created tensions with Khartoum, which, with Beijing's support, has complicated the process of UN force deployments in the region. According to one Chadian diplomat, "As China has a veto (on the UN Security Council) we have to have it by our side so the draft (UN resolution) on the presence of UN forces (in Darfur) goes through without difficulty."[36] Beijing's inducements included economic and infrastructure packages, as well as a promise to reduce both its own and Sudan's support for Chad's rebels.[37] These methods are consistent with Beijing's use of financial and political carrots and sticks in its relations with African nations.

Energy and Natural Resources

China's demand for energy is at the forefront of its Africa strategy. Despite its vast land mass, the PRC only has a small percentage of the world's known oil reserves.[38] An oil exporter until the early 1990s, China imported 3,384 thousand barrels of oil per day in 2005—just under half of its total daily consumption of 6,988 thousand barrels of oil per day.[39] To meet its citizens' surging energy needs, and reduce dependence on environmentally hazardous and inefficient coal, China has been searching abroad for petroleum resources. About a quarter of these imports come from African nations, including Sudan, Chad, Libya, Nigeria, Algeria, Gabon, and Angola.[40]

China's approach to securing energy supplies differs from that of the Western countries. Beijing seeks to gain control of African oil at its source to ensure supplies and help insulate its economy from price shocks. "China's energy policy is based upon a 'strategic' approach which eschews dependence on markets."[41] As a result, China's crude supplies are heavily dependent on bilateral ties to oil-producing states. Beijing's cultivation of relationships with African elites facilitates its oil companies' exploring, securing, extracting, processing, and shipping African crude.

To mitigate China's dependence on Middle Eastern supplies, Chinese companies are aggressively seeking to expand their investments in Africa's energy industry. Africa is a particularly attractive destination for Chinese energy companies because the continent has a sizable share of the world's most desirable sweet crude. In 2004, China displaced Japan as the second largest importer of African oil, after the United States.[42] According to statistics from the multinational British Petroleum (BP), Chinese imports of African oil increased more than 71 percent between 2003 and 2005.

Although Beijing's energy purchases from each African country make up only a small percentage of total Chinese imports, they represent a significant share of the African oil producers' exports. For instance, while Sudan accounts for roughly 6 percentage of total Chinese crude imports,[43] roughly 50 percent of Khartoum's oil exports go to China. Beijing imports a quarter of Angola's oil and an increasing percentage of exports from Equatorial Guinea, Nigeria, and Gabon.[44] These are poor countries and petroleum exports account for a sizable part of their GDP. As such, the effect of China's energy purchases on these countries' domestic political and social development is noteworthy.

Beijing offers a range of incentive packages to woo resource-rich African nations. It has offered loans and grants, debt relief, weapon sales, infrastructure projects, personnel training, and student scholarships, and approved a variety of country-specific requests. Angola, for example, has begun to reap the rewards. Beijing has financed the construction of new roads, modern hospitals, and a new Ministry of Justice building, and supplied millions of dollars in loans and financing to the oil-rich nation.[45] In 2003, Angola increased crude sales to China by 77 percent, replacing Oman as China's third largest oil supplier behind Saudi Arabia and Iran.[46]

Nigeria has also benefited. In January 2006, China's state-controlled oil firm, CNOOC, agreed to pay $2.3 billion for a 45-percent stake in a Nigerian offshore oil field, and in April of that year China committed $4 billion in oil and infrastructure investment projects. Beijing also agreed to purchase a controlling stake in Nigeria's 110,000 barrel-per-day (bpd) Kaduna oil refinery, build various infrastructure projects including a railroad system and power stations, and provide telecommunications and satellite technology.[47]

Beijing's moves have not come without local resistance. In Nigeria, militants in the oil-rich Niger Delta have responded to Beijing's purchases by warning that investing in "stolen crude" places Chinese citizens in the "line of fire."[48] Nigeria's government, after months of unsuccessfully pressuring Western nations for arms, has now turned to China for the naval patrol boats and arms it needs to help protect Niger Delta rigs from rebel attacks.[49]

Sudan's oil is particularly attractive because its low sulfur content is well-suited to Chinese refineries. Sanctions imposed by Western nations also make this nation a more willing economic and political partner for Beijing. China National Petroleum Corporation (CNPC) holds the largest share (40 percent) in Sudan's largest oil production project, the Greater Nile Petroleum Operating Company. In 2005 alone, China imported about 200,000 bpd from Sudan,[50] and increases are expected as a result of a peace and revenue-sharing agreement that should allow new fields to come online in previously war-torn areas.[51] Oil revenues have gone directly to the Sudanese government, which in turn has purchased arms and other military equipment from suppliers, including China. Looking forward, Sudan's leaders are likely to continue relying on Chinese oil firms for lucrative deals as long as Western firms are restricted by sanctions and deterred by Khartoum's human rights violations.

China's resource interests in Africa extend beyond energy. Beijing imports mineral resources, steel, and cotton from the continent.[52] China is now the world's largest copper consumer, and in 2005, China consumed roughly one-third of the total global output of steel, 40 percent of cement, 26 percent of copper, and its share of the world's energy consumption has risen from 9 to

12 percent in a decade.[53] Beijing has made sizable purchases of copper from African producers like the Democratic Republic of Congo, and in Zambia, Chinese copper investors provided the lion's share of the $100 million in Chinese investment between 2001 and June 2005.[54]

Zimbabwe is attractive to China because it has the second largest deposits of platinum in the world as well as more than 40 other minerals, including ferrochrome, gold, silver, copper, and uranium.[55] Zimbabwe's leaders are aware of China's race to find raw materials for its economic growth, and they use their country's natural resources as a lure. The governor of the Reserve Bank of Zimbabwe, Dr. Gideon Gono, said in a May 2005 meeting with the deputy governor of the People's Bank of China: "I would like to unveil to the Chinese people the vast investment opportunities that abound in Zimbabwe, including our natural resource endowments."[56]

Beijing has secured coal and coke concessions and is also working to gain access to Harare's substantial uranium deposits. China's National Aero-Technology Import and Export Corporation (CATIC) and China North Industries Corporation (NORINCO), for instance, have agreed to finance multi-billion dollar expansion projects by Zimbabwe's Electricity Supply Authority and Hwange Colliery Company to build power generators and supply mining equipment in return for coal concessions.[57] In May 2006, a team of Chinese scientists visited Makuti in the Zambezi Valley in what could be the beginning of extensive Chinese uranium exploration in Zimbabwe.[58]

In the case of Zimbabwe, however, there are signs the commercial relationship may be strained. In May 2006, for instance, a half dozen state-run Chinese firms abandoned construction sites and froze operations because of nonpayment.[59] Harare also defaulted on a scheduled $12 million debt repayment to China's state-owned Aviation Industry Corporation of China (Avic) which had supplied Air Zimbabwe with two MA60 passenger planes under a credit arrangement.[60]

Lastly, Africa is a primary source for Chinese demand for illegal timber and ivory. In the wake of the deadly floods along China's Yangtze River in 1998, Beijing imposed domestic logging restrictions. As a result, Chinese firms must look overseas for timber, and have increased imports from Africa and other suppliers in the developing world. Between 1998 and 2005, Chinese timber imports from African nations such as Liberia and Gabon surged, and with them China's role in the black-market timber trade grew. The secretive nature of black-market timber and illegal ivory imports make them impossible to accurately calculate, but the availability of both resources on the Chinese market and expert testimony confirm a robust trade.[61]

Table 2.1

China's Top Trade Destinations in Africa (in billions of US$)*

	2002 total trade	Year-on-year % change	2003 total trade	Year-on-year % change	2004 total trade	Year-on-year % change	Jan.–Aug. 2005 trade	Year-on-year % change (Jan.–Aug.)
South Africa	2.58	16.1	3.871	50.1	5.912	52.8	4.696	25.6
Angola	N/A	N/A	N/A	N/A	4.910	108.8	4.276	73.1
Sudan	1.55	33.8	1.92	23.9	2.521	31.3	2.430	55.6
Nigeria	1.168	2.1	1.859	59.1	2.182	17.5	1.943	41.7
Egypt	0.945	–0.1	1.09	15.4	1.576	44.7	N/A	N/A

*Many experts contend that Chinese customs statistics are unreliable. Despite their questionable accuracy, however, they are useful in that they reveal the trend of the rapid growth in China's trade relations with Africa. For the direction of trade statistics on the region as a whole, please see Appendix I. It is also worth noting that after October 2005, Xinhua stopped publication of China's trade with the major African nations.

See "China's Trade with Major African Partners," Xinhua (February 1, 2003); (February 22, 2004); (February 9, 2005); (October 22, 2005).

Commercial Interests

Two-way trade between China and Africa rose from $10.6 billion in 2000 to $40 billion in 2005 and continues to increase, according to Chinese government statistics.[62] Sino-African trade is balanced. However, with the exception of trade with oil-rich African nations such as Angola, Sudan, and Nigeria, Beijing's exports dominate its bilateral trade relationships in Africa (see Table 2.1). Chinese producers have secured large portions of the African textile and electronics markets and have established their own sales outlets throughout the continent. As with other parts of the world, however, the rapid inroads made by Chinese producers have also increased tensions.

In the Horn of Africa in 2004, Chinese exports to Ethiopia included over 93 percent of the bilateral trade. In the first half of 2005, Chinese purchases from Djibouti, Eritrea, and Somalia/Somaliland were negligible.[63] Attempting to correct the lopsided trade relationship, Beijing scrapped tariffs on 190 commodities from twenty-five Africans nations in 2005. Given the nature of Chinese tastes and preferences, however, this decision is unlikely to dramatically change China's trade relationships in the region.[64]

In many African countries, concern about growing trade deficits, export competition, and poor quality Chinese goods is growing. In Kenya, for instance, textile exports to the United States in the first quarter of 2005 dropped 13 percent compared to the first quarter of 2004.[65] The rapid drop off was largely seen as the result of increased U.S. imports from China, and led to thousands of layoffs in the east African country. South Africa and Lesotho have also seen tens of thousands of workers in the textile sector lose their jobs, and there is a growing backlash against Chinese products and workers.[66] In Nigeria, concerns about counterfeit Chinese pharmaceuticals led officials in Lagos to ban dozens of Chinese and Indian firms' products in January 2006.[67]

Nevertheless, China continues its efforts to secure more predictable market access to Africa. In November 2005, China's newly appointed special representative to the Common Market for Eastern and Southern Africa (COMESA) met with Secretary General Erastus J.O. Mwench in Lusaka, Zambia.[68] COMESA—which includes twenty African nations[69]—and China have been working toward a free trade agreement (FTA). Secretary General Mwench commented on the agreement's underpinnings: "China is a major consumer of some raw materials we found in COMESA, like copper and oil. And we believe China has got the technology that we could be looking at, like in telecommunications, industry and agribusiness."[70] This agreement promises to increase trade volume and coordination between China and COMESA's member countries, especially natural resource exporters with close ties to Beijing.[71]

China and the Southern African Customs Union (SACU) have also begun negotiations for a FTA. Beijing is strongly in favor of the agreement, which would help secure raw materials. SACU, which includes Botswana, Lesotho, Namibia, South Africa, and Swaziland, remains hesitant because of fears of reduced tariff revenue and cheap Chinese imports undermining local production. Negotiations, which began in October 2004, are also complicated by Swaziland's and Lesotho's close dealings with Taiwan. Swaziland maintains formal diplomatic relations with Taiwan and is home to $50 million in Taiwanese investment. In 2004, total Taiwanese investment in Lesotho stood at approximately $600 million, with the island employing around 20 percent of Lesotho's workforce. As negotiations progress, Beijing is expected to pressure SACU to urge Swaziland to switch its recognition to Beijing.[72]

China is also eager to sign an FTA with South Africa, which represents roughly 20 percent of total Chinese trade on the continent.[73] The bilateral trade relationship is already growing fast. Between 2003 and 2004, trade jumped 53 percent, and between 2004 and 2005 bilateral trade increased 23 percent—from $6 billion to $7.3 billion.[74] But again, however, South African concerns over an influx of cheap Chinese consumer goods, competition for export markets, and a growing trade deficit have caused tensions and threaten to undermine the commercial and political relationship if left unchecked. During an official state visit to South Africa in June 2006, Wen Jiabao tried to alleviate these tensions by calling for "self-restrictive measures" such as export licenses and a self-imposed maximum annual growth rate on Chinese textile exports.[75]

Methods

To achieve the objectives described above, Beijing employs methods that can be grouped into the following categories: China's "soft approach," agricultural and technical assistance, investment and economic support, military patronage and peacekeeping, and diplomatic support. Taken together, these methods have helped Beijing secure its interests in African capitals. They have also, however, caused some backlash in African nations where Chinese exports—whether they be low-priced electronics or arms—are perceived as contrary to the interests of the local population.

China's Soft Approach

China's soft approach refers to humanitarian, cultural, and educational projects aimed at promoting China's image and influence. China's use of charm or "soft influence" in Africa dates back to the 1950s and 1960s, when its agricultural, medical, and military training programs first began. While Beijing's methods may be rooted in the past, today's programs' objectives have changed. Although they are still designed to improve China's image and develop person-to-person links, unlike the ideology-driven campaigns of the past, the primary objective is now to counter adverse feelings generated by China's commercial and political expansion on the continent.

Every year, cities throughout China and Africa host gatherings designed to foster goodwill and personal ties. In August 2004, Guangzhou hosted the China-Africa Youth Festival and, as mentioned, Beijing hosted the third ministerial meeting and first Summit of the China-Africa Cooperation Forum in November 2006. Meetings like these provide a venue to roll out Beijing's technical assistance and support initiatives. Following the first China-Africa Cooperation Forum in 2000, Beijing set up the African Human Resources Development Fund to provide personnel training to 10,000 Africans.[76] This fund has sponsored over three hundred courses and trained hundreds of African personnel in areas of diplomacy, economic management, national defense, agriculture, education, science, technology, and medical treatment.[77]

Medical cooperation is another traditional form of goodwill building. China deployed its first medical team to Africa in 1964, at the invitation of the Algerian government. Since the 1960s, China has helped forty-seven African nations develop their health services, and over 15,000 Chinese doctors have treated roughly 180 million African patients. Many of these programs remain, or have been resurrected with direct government support or via the China-Africa Cooperation Forum.[78] According to 2003 statistics, Beijing had 860 medical personnel in thirty-five teams in thirty-four countries in Africa.[79]

Chinese medical schools and physicians also train African doctors, and provide medicine and equipment free of charge to African countries. In 2002, the Chinese Ministry of Health conducted a two-part international training course in techniques for the prevention and treatment of malaria and tropical diseases, in which students from seventeen African countries participated. That same year, as part of the China-Africa Cooperation Forum, China and twenty-one African countries convened a Sino-African forum on traditional medicine and pharmaceuticals.[80]

China's ability to project goodwill through its health initiatives in Africa, however, will likely diminish as hundreds of millions of Chinese demand better health services. This, coupled with China's fast improving medical facilities, means China's contribution in this area will increasingly be comprised of medical technology and medications, as opposed to doctors administering treatment directly to Africans as has traditionally been the case. To combat malaria, for instance, the United States-based PATH Malaria Vaccine Initiative and the Chinese company Shanghai Wanxing Bio-Pharmaceuticals are jointly developing a vaccine.[81]

Tourism is another element in China's soft approach. Beijing is encouraging Chinese tourism in Africa, with an eye toward developing cultural and business ties. Memoranda of Understanding on tourism have already been signed between China and Ethiopia, Mauritius, Kenya, Seychelles, Tanzania, Zimbabwe, and Tunisia.[82] In 2003, China awarded its Approved Destination Status to eight African nations, including Kenya, Zimbabwe, and South Africa.[83] South Africa is Chinese tourists' favorite African destination, and between 2003 and 2004 Chinese tourists tripled.[84] In 2004–5, however, the murder of twenty-four Chinese citizens in South Africa made headlines in China and caused a nearly 15 percent drop in 2005.[85]

Bringing African students to China to study is another approach Beijing uses to project a positive image in Africa. Between 1956 and 2005, China awarded approximately 15,300 government scholarships to students from fifty-two African nations. In 2003, 1,793 African students studied in China, representing one-third of the government-sponsored foreign students that year.[86] Beijing also seeks to establish "Confucius Institutes" in Africa—programs at leading local universities, funded by Beijing and devoted to the study of Chinese culture and language training. Already in Asia, Confucius Institutes have proved effective in encouraging graduate students to focus on the study of China, and ultimately, to actually study in China. As with other cultural initiatives, China's scholarship programs develop trust by investing in long-term relationships with

African elites. These educational initiatives help provide China with the kind of workforce it requires to expand its own high-tech industries in Africa.[87]

Agricultural and Technical Assistance

Technological advancements in China, particularly in the field of agriculture, have provided Beijing with another powerful enticement for poor African countries. Agricultural cooperation and technical assistance are an indication of the trajectory of Chinese power projection in Africa. They cost comparably little, but effectively promote Beijing's economic development model to African countries and improve China's image throughout the continent. China has already transferred technology to African nations in areas such as engineering, mechanics, agriculture, and telecommunications and over 10,000 Chinese agricultural experts have established nearly two hundred projects including farms, agro-technical stations, and agricultural technician training programs throughout Africa.[88]

China's efforts include agro-processing projects in Namibia and cotton production in Zambia.[89] Beijing supplies Zimbabwe with expertise, technical assistance, and agricultural equipment, including tractors and agroprocessing.[90] Even Zimbabwean President Robert Mugabe's ruinous redistribution of land and destruction of his country's agricultural resources have proved to be an opportunity for China's businesses. Not only have Chinese companies worked closely with President Mugabe's regime, but according to Chinese Premier Wen Jiabao, "China respects and supports efforts by Zimbabwe to bring about social justice through land reform."[91]

Even in the wealthier and more developed South Africa, Chinese technology and scientific transfers are increasing. In March 2003, at the Second Session of the Science and Technology Joint Committee between China and South Africa in Pretoria, Wu Zhongze, Chinese vice minister of Science and Technology, and his South African counterpart signed an agreement launching technical and scientific cooperation in ten major projects. These included "peaceful use of nuclear energy," mini-satellites, remote sensing, mining, space, medicine, and health care.[92]

Beijing is also working with international institutions to provide technical assistance. In May 2004, for example, Shanghai hosted the Conference on Scaling up Poverty Reduction. Over 1,000 delegates attended the two-day event, co-sponsored by the World Bank and the Chinese government, which featured keynote speeches by Premier Wen Jiabao, World Bank President James Wolfensohn, and Uganda's President Yoweri Museveni. The Conference's numerous sessions on the promotion of agricultural and technical development highlighted China's desire to lead and support the developing world in these areas. By holding the conference in Shanghai, arguably the nation's most impressive city, China bolstered its prestige among conference participants.[93]

Beijing helps African regimes craft their images at home through propaganda and information control, and by providing technology and training for press and Internet monitors. On November 11, 2005 China's official press even alluded to Beijing's methods, claiming "In the information sector, China has trained dozens of media for 35 African countries for the past two years."[94] In a noteworthy coincidence, the day before this statement, Reporters Without Borders released its analysis: "The use of Chinese technology in a totally hypocritical and non-transparent fashion reveals the government's iron resolve to abolish freedom of opinion in Zimbabwe."[95]

Investment and Economic Support

One of China's primary contributions to Africa's economic development comes in the form of investment. Between 1956 and 2005, China provided aid to more than 800 infrastructure projects in Africa.[96] Believing that the more visible the project, the more goodwill it generated, early Chinese projects were additions to Africa's landscape, such as sports stadiums built in Somalia, Sierra Leone, Gambia, and the Democratic Republic of the Congo.[97] Building has continued, and in 2006 China offered to construct another stadium—this one in Guinea—and has built the Ministry of Foreign Affairs buildings for Djibouti, Eritrea, and Uganda.[98] Beijing also finances the construction of roads, schools, and housing throughout Africa. In 2003 and 2004, China funded several government buildings in Gabon, an Algerian airport terminal, and Rwanda's new convention center and Ministry of Foreign Affairs building.[99] Increasingly, however, Chinese investment reflects the country's growing entrepreneurial spirit and need to sustain domestic economic development.

Chinese investors, many working for state-run firms, have launched an aggressive campaign in Africa. As part of this effort, the Chinese government had set up eleven centers for "investment and trade promotion" by 2004, providing business and consultation to Chinese enterprises in Africa.[100] Beijing has also established special funds, simplified procedures, and put supporting policies in place to further encourage Chinese investment in Africa. Chinese businessmen are particularly interested in resource exploitation, infrastructure construction, telecommunications, and the African pharmaceutical industry.[101] Through these investment programs, China has become one of Africa's most important partners and has extended its network of purchasers and suppliers. Investment projects, many of them on commercial terms, extend China's economic reach and influence.

The first Sino-African business conference was held in December 2003 and resulted in the signing of agreements on twenty projects with a total value of $680 million.[102] In all, China has struck deals with over forty African countries. The Chinese Telecom firm, Huawei, holds contracts worth $400 million to provide mobile phone service in Kenya, Zimbabwe, and Nigeria. In Zambia, Chinese investors are working on a $600 million hydroelectric plant at Kafue Gorge and are also active in South Africa's and Botswana's hotel and construction industries.[103]

Sudan and Ethiopia have been among the largest beneficiaries of Chinese investment. As of January 2006, Chinese investment in Sudan alone totaled $4 billion, making it the country's largest investor.[104] Using imported Chinese laborers, companies have constructed roads and bridges, power and water supply stations, irrigation and telecommunications networks, and housing in east Africa. Chinese construction firms have been known to submit bids below cost in an effort to break into markets. Specific examples of Chinese projects include the Oratta Hospital in Asmara, the Addis Ababa circumferential highway, and the $300 million hydroelectric dam and power plant on Ethiopia's Tekeze River.[105]

Beijing also encourages economic development through grants, debt relief, and low-cost loans to African countries. In 2002, China gave $1.8 billion in development aid to its African allies.[106] Examples of Chinese gifts include a $1.3 million "technical assistance" grant to Kenya in 1996,[107] $6 million and $2.5 million grants to Zambia and Nigeria in 2003, respectively, [108] and a $3.72 million grant to Niger in December 2005.[109] While these totals may seem small by Western standards, they buy influence in poor African nations.

In recent years, China has used debt relief to assist African nations, effectively turning loans into grants. Since 2000, Beijing has cancelled the debts of thirty-one African countries. During the first China-Africa Cooperation Forum in 2000, China reduced or wrote off $1.2 billion in debts owed by African nations.[110] Again in 2003, during the lead up to the second China-Africa Cooperation Forum, Beijing granted another round of relief—this time totaling $750 million.[111] Ethiopian Prime Minister Meles Zenawi explained the importance of debt relief to China's relations with African nations: "China's exemplary endeavor to ease African countries' debt problem is indeed a true expression of solidarity and commitment."[112] China's willingness to provide debt relief has endeared it to Africans while pressuring Western countries to begin similar programs. China, however, also uses this tool as leverage. In 2005, Beijing granted Senegal an $18.5 million reprieve, but only after the African nation broke diplomatic ties with Taiwan.[113]

In some nations, however, China's loans and aid packages undermine attempts to improve government transparency and accountability. In Angola, for example, a majority of the country's roughly 13 million people live in poverty, and elites have siphoned off much of the nation's oil wealth. As part of a larger package in March 2004, China provided Luanda over $2 billion in loans in accordance with its principle of noninterference in other countries' internal affairs. This move allowed Angola to turn its back on an International Monetary Fund package that would have demanded greater accountability and the adoption of improved governance measures. In November 2005, Jose Pedro de Morais, Angola's finance minister, said he expected future Chinese loans would exceed $2 billion. He added, "When we ask our Chinese counterparts if they are willing to provide more loans, they say yes."[114] But in Angola, as in many other petroleum producer countries, Beijing's loans are backed by oil—if an African country cannot repay loans in cash, Beijing will take oil. Loans are also often targeted at infrastructure projects that facilitate development of the petroleum industry.

Beijing's methods have matured as China's economic system has developed and demonstrate the increased sophistication with which China's leaders handle commercial relations in Africa. The lure of the Chinese market and increased trade and investment flows are used to garner support in Africa, and Chinese demand for African raw materials and minerals has pushed up prices to the delight of many powerful Africans. In 2006, for instance, China's quest for oil allowed Nigeria to squeeze $4 billion in infrastructure and health investment from Beijing in exchange for drilling licenses, $2.3 billion for a 45 percent stake in an offshore oilfield, an $800 million deal to provide 30,000 barrels of oil a day, and an agreement to launch Nigeria's first space satellite.[115]

Military Patronage and Peacekeeping

One of the most intriguing aspects of China's relations with African countries is military cooperation, sales, and assistance. Although the clandestine nature of these operations makes them most difficult to accurately document, Chinese firms are among the top suppliers of conventional arms on the African continent. Between 1996 and 2003, Chinese arms sales to Africa were second only to Russia's, making up roughly 14 percent ($900 million) of Africa's total conventional arms imports.[116] Beijing openly discusses the use of military patronage in its January 2006 Africa policy white paper:

> China will promote high-level military exchanges between the two sides and actively carry out military-related technological exchanges and cooperation. It will continue to help train African military personnel and support defense and army building of African countries for their own security.

In an effort to increase its influence throughout the continent, Beijing has expanded military-to-military ties and arms sales to dozens of African governments.[117] These ties are not generally intended to bolster China's military strength, but rather to court resource-rich African nations and protect Chinese commercial interests.

To secure its state-run companies' extensive oil investments, Beijing has established its most extensive military ties with Sudan. Beijing began selling arms to Khartoum in the 1980s and continues to be Sudan's largest supplier of arms. Beijing supplied both small and sophisticated arms to the regime during the two-decade-long north-south civil war, and these arms remain critical to Khartoum's armed forces today.[118]

In January 2006, a Council on Foreign Relations task force summed up China's military sales to Sudan:[119]

> Chinese weapons deliveries to Sudan have included ammunition, small arms, towed howitzers, anti-aircraft guns, anti-personnel and anti-tank mines, tanks, helicopters, and fighter aircraft. China also helped establish three weapons factories in Sudan, including one for assembling T-55 tanks. There are also an undetermined number of Chinese military personnel stationed in Sudan to secure its investments.[120]

Concerns about genocide and civil war in Sudan's Darfur region have not led Beijing to reduce its military or diplomatic support for Khartoum. In fact, in November 2005, amid widespread international condemnation of genocide in Darfur, the two countries' long-standing military ties were renewed when Khartoum's deputy chief of general staff Muhammad Isma'il visited Beijing and met with Vice Chairman of China's Central Military Commission Xu Caihou to "increase military exchanges and cooperation."[121]

Beijing also has close military ties with Zimbabwe. In April 2005, Zimbabwe's air force received six K8 jet aircraft to be used for training jet fighter pilots and for "low intensity" military operations.[122] The year before, a Chinese radar system was installed at President Mugabe's mansion in the Harare suburbs.[123] But most importantly, Zimbabwe's defense ministry confirmed the purchase of 12 FC-1 fighter planes and 100 military vehicles, worth an estimated $240 million, in June 2004.[124] This order, which had been kept secret, circumvented the state procurement board that had been tasked with appropriating Zimbabwe's $136 million defense budget.[125]

Beijing sells large quantities of arms in East Africa, as well. China was once an important source for Somalia's weaponry and from 1998 to 2000, while Ethiopia and Eritrea were embroiled in conflict, Chinese firms shipped large amounts of arms to both parties.[126] Even as UN Security Council members, including China, expressed fears of a new conflict between Ethiopia and Eritrea in the fall of 2005, China was deepening military cooperation and extending arms sales to both.[127] Indeed, both Beijing and Moscow supplied weapons while at the same time supporting UN resolutions calling on the parties not to use them. One observer called this situation "a windfall for arms merchants on the so-called 'security' council."[128]

Despite its policy of noninterference in foreign countries' international affairs, Chinese-made weapons have ended up in the hands of African rebel groups. Chadian rebels, for instance, who tried to overthrow the government in April 2006 used Chinese-made AK-47 rifles, rocket-propelled grenades, and more advanced artillery pieces mounted on SUVs that reportedly came via Sudan. According to Amnesty International, China has also aided rebel groups in Angola and the Democratic Republic of Congo.[129] China's military sales in Africa are troubling, but there is, however, another aspect of China's security presence in Africa that holds promise. In 2001, Beijing began contributing large numbers of peacekeepers to United Nations' missions in Africa. Hundreds of Chinese peacekeepers are now in the Democratic Republic of Congo and Liberia, and more are gearing up to help in Sudan. According to China's defense ministry, the Chinese government sent 550 peacekeeping troops to Liberia in 2004, including a 240-member transport company, a 275-member engineering team, and thirty-five medical staff for a UN hospital.[130] Integrating Chinese forces into UN efforts to mitigate conflict in Africa is an effective way to build military-to-military ties, political capital in Western nations, and credibility on a local level.

Diplomatic Support

China continues to support its African allies' initiatives in international organizations. Beijing has supported UN Security Council resolutions to secure aid for Liberia and softened resolutions condemning Khartoum for the genocide in Darfur. China provided the loans necessary for an impoverished Angola to snub the International Monetary Fund, and Chinese support for Chad has strengthened the nation's embattled leadership in its dealings with the World Bank. China has also supported proposals favored by African nations on UN Security Council reform, peacekeeping, and debt relief. In this way, China secures support in African capitals and underscores its support for the developing world.

Premier Wen Jiabao alluded to this method:

> As a permanent member of the UN Security Council, China will always stand side-by-side with developing countries in Africa and other

> parts of the world, and support their legitimate requests and reasonable propositions.[131]

In addition to Sino-African cooperation in international institutions, at least one high-level delegation of African leaders visits Beijing nearly every month. Alongside struggling democracies, Beijing hosts dictators and despots.[132] In 2005, China came under international scrutiny for its diplomatic support for Sudan and Zimbabwe. In particular, Beijing's hosting and praise of Zimbabwean strongman Robert Mugabe and attempts to water down UN resolutions targeting Khartoum were criticized in Western capitals. Yet this pressure had little apparent effect on either bilateral relationship. China and Zimbabwe signed a trade deal, but refused to disclose the contents, and China has reaffirmed its steadfast support for the Khartoum regime.[133]

Conclusion

Since the end of the Cold War, a largely indifferent Russia and a United States preoccupied with engagements stretching from the Middle East to Northeast Asia have afforded China an historic opportunity to engage Africa. As a result, China has become, in the words of Zimbabwe's strongman Robert Mugabe, "an alternative global power point" in Africa.[134] Beijing's strategy in Africa has been designed to ensure access to energy and other critical natural resources, open new markets for exporters, and enhance China's ability to safeguard its interests in international fora and institutions. In pursuit of these objectives, Beijing has been pragmatic and has shown little regard for ideology or humanitarian constraints that might give pause to leaders that must answer to democratic domestic polities.

In the years to come, Africa's importance to China will grow apace with China's need for raw materials and its claims that it leads the developing world. Chinese investment and trade have helped many African countries develop industries, unearth resources, and increase the value of African exports. Unfortunately, Chinese economic, diplomatic, and security policies turn a blind eye to corruption and poor governance. But Beijing's policies and investment decisions in Africa may also have unintended long-term consequences for China. In Gabon and Zimbabwe, for example, China's lucrative relationships with aging presidents-for-life could make bilateral ties with these countries difficult to manage during the uncertain power transitions looming on the horizon. Meanwhile, arms sales such as those in the Horn of Africa could undermine a growing market for Chinese consumer goods.

For the United States, conflict prevention and resolution, HIV/AIDS, human rights, building democracy, and the search for non-Gulf oil sources are likely to remain priorities in Africa. Neither Beijing nor Washington wants Africa to become a region of aggressive competition between the two sides. However, given competing interests for energy and other natural resources, and divergent views on democratization and human rights, China's African strategy will require creative and nuanced responses from American policymakers. Although the long-term implications of China's proactive strategy in Africa remain in doubt, Beijing's reach into Africa will almost certainly deepen. As China expands political, economic, and military ties throughout the continent, Sino-African relations will continue to provide opportunities and challenges for the African nations, China, and the global community.

Notes

1. W.A.C. Adie, "Chinese Policy Towards Africa," in *The Soviet Bloc, China and Africa,* ed. Sven Hamrell and Carl Gosta Widstrand (Ussala: The Scandinavian Institute of African Studies, 1964), 44.

2. Bruce D. Larkin, *China in Africa 1949–1970: The Foreign Policy of the People's Republic of China* (Berkeley, CA: University of California Press, 1971), 157.

3. Adie, "Chinese Policy Towards Africa," 53.

4. Larkin, *China in Africa 1949–1970*, 156.

5. New China News Agency (November 8, 1960), in Adie, "Chinese Policy Towards Africa," 53.

6. Patrick Tyler, *A Great Wall* (New York: Public Affairs, 1999), 204.

7. James Lilley and Jeffrey Lilley, *China Hands: Nine Decades of Adventure, Espionage, and Diplomacy in Asia* (New York: Public Affairs, 2004), 146.

8. Ibid., 155.

9. S. S. Kim, "China and the Third World: In Search of a Neorealist World Policy," in *China and the World: Chinese Foreign Policy in the Post–Cold War Era,* ed. S.S. Kim (Boulder: Westview Press, 1984), 184.

10. Richard Lowenthal, "The Sino-Soviet Split and its Repercussions in Africa," in *The Soviet Bloc, China and Africa,* ed. Sven Hamrell and Carl Gosta Widstrand (Ussala: The Scandinavian Institute of African Studies, 1964), 132.

11. John A. Marcum, "Lessons from Angola," *Foreign Affairs* 54, no. 3 (April 1976): 410–413.

12. Eugene K. Lawson, "China's Policy in Ethiopia and Angola," in *Soviet and Chinese Aid to African Nations,* ed. Warren Weinstein and Thomas H. Henriksen (New York: Praeger, 1980), 174.

13. Ibid., 175.

14. "Central Organs Announce Mao's Passing," Xinhua (September 9, 1976), in *Chinese Politics from Mao to Deng,* ed. Victor C. Falkenheim and llpyong J. Kim (New York: Paragon, 1989), 28.

15. Lawson, "China's Policy in Ethiopia and Angola," 172.

16. An example of this type of rhetoric appears on China's Ministry of Foreign Affairs Web site: "Close cooperation has been established in international forums while handling the issue of human rights and other important issues. In those forums they [Chinese and African leaders] made joint efforts to maintain the lawful rights of developing countries and push forward the creation of a new, fair and just politi-

cal and economic order in the world." "China-Africa Relations" (posted on China's Ministry of Foreign Affairs official Web site, undated). Available at www3.fmprc.gov.cn/eng/29343.html.

17. Yuan Ye, "China, S. Africa Consolidate Ties for Developing World's Benefit," Xinhua (June 20, 2006).

18. "Full text of Hu Jintao's Speech at Asian-African Business Summit Reception." Available on the China-Africa Cooperation Forum Web site at www.fmprc.gov.cn/zflt/eng/zxxx/t192739.htm, downloaded on April 22, 2005.

19. Willy Wo-Lap Lam, "Africa Trip Boosts China's Status." Available on CNN.com, downloaded on December 17, 2003.

20. "Hegemony is Cold-Shouldered by Many People: Comment," *People's Daily* (June 29, 2005). Available on http://english.people.com.cn/200506/29/eng20050629_193109.html.

21. "Text of Wen Jiabao's Speech at China-Africa Cooperation Forum," Xinhua Domestic Service [Chinese] (December 15, 2003).

22. Tarun Chhabra, "Opening a Sino-U.S. Dialogue on Africa," *Center for Strategic and International Studies Prospectus Papers*, vol. 2, no. 3 (2001). Available at www.csis.org/pubs/prospectus/01fall_chhabra.htm.

23. "China to Double Aid to Africa," Integrated Regional Information Networks, United Nations, November 6, 2006. Available at www.worldpress.org/Africa/2554.cfm.

24. "China: Chinese Diplomacy, Western Hypocrisy and the U.N. Human Rights Commission," *Human Rights Watch,* vol. 9, no. 3 (March 1997): 7. Available at www.hrw.org/reports/1997/china2/China-03.htm#P100_21248.

25. "Profit and Prejudice: China in Africa," *China News Analysis*, no. 1574 (December 15, 1996), in ibid.

26. Technically, the ROC on Taiwan still claims to be the rightful sovereign of all China, according to its constitution. In 1991, President Lee Teng-hui acknowledged the reality that they do not actually control the mainland today, but still technically the figleaf of "rightful" ROC sovereignty remains.

27. "Ethiopia's Parliament Backs China's Anti-Secession Law," Xinhua (March 15, 2005).

28. Howard French, "Taiwan Competes with China to Win African Hearts," *The New York Times* (January 24, 1996): 3.

29. "Chad Chooses China over Taiwan," *BBC* (August 7, 2006).

30. "Taiwan-Africa Relations: A New Diplomatic Approach?" *China Post* (January 10, 2003).

31. "Introduction" (posted on Africa-Taiwan Economic Forum official Web site, undated). Available at www.africa.org.tw/introduction_english.asp.

32. Still, the traditional competition between Taipei and Beijing has not entirely subsided. During his world tour in May 2006, President Chen made a surprise visit to Libya. A diplomatic row had erupted a few months prior, when Tripoli reportedly extended diplomatic recognition to Taipei. Tripoli and Taipei had apparently agreed to exchange representative offices when Chinese Foreign Minister Li Zhaoxing arrived on a two-day official visit to Libya and was reassured by Libyan Foreign Minister Mohammed Abdel-Rahman Shalgam that Tripoli "only recognizes one-China." "Libya Says It Only Recognizes One-China," Jana [official Libyan news agency] (January 19, 2006).

33. William Foreman, "Liberia Cuts Diplomatic Ties with Taiwan, Officials Say," Associated Press (October 12, 2003).

34. "Taiwan's Foreign Minister Offers Resignation Over Liberia Setback," Deutsche Presse-Agentur (October 13, 2003).

35. "Chinese Republic Promises 600 Peace-keeping Force for Liberia," *The Perspective* (January 20, 2004). "China Offers Liberian Farmers $1 Million in Farm Tools," Xinhua (May 5, 2006).

36. "Chad Chooses Realism, Cash and China over Taiwan," Agence France-Presse (August 6, 2006).

37. Ibid.

38. *BP Statistical Review of World Energy* (London: British Petroleum Company, 1998), 4. Available in "Energy Demand and Supply in China," at www.rand.org/publications/MR/MR1244/MR1244.ch2.pdf.

39. *BP Statistical Review of World Energy, 2004–2006.*

40. David H. Shinn, "China's Approach to East, North and the Horn of Africa" (Testimony before the U.S.-China Economic and Security Review Commission, July 21, 2005). Available at www.uscc.gov/hearings/2005hearings/written_testimonies/05_07_21_22wrts/shinn_david_wrts.htm.

41. Philip Andrews-Speed, "State Control is the Cause of China's Energy Crisis," *Wall Street Journal* (April 30, 2004).

42. "Forget Mao, Let's Do Business," *The Economist* (February 7, 2004).

43. Jon D. Markman, "How China Is Winning the Oil Race," *MSN: Money Central* (April 26, 2006).

44. Lindsay Hilsum, "China's Big Investment," *The Online NewsHour with Jim Lehrer* (July 5, 2005). Available at www.pbs.org/newshour/bb/asia/july-dec05/china_7-05.html.

45. "Chinese Credit Woos Angolan Oil," Reuters (February 22, 2005). Available at http://edition.cnn.com/2005/BUSINESS/02/21/china.angola.reut/.

46. Lawrence Yong, "African Oil Exports to China Soar," *International Herald Tribune* (May 6, 2004).

47. "China and Nigeria Agree Oil Deal," *BBC (*April 26, 2006).

48. Craig Timberg, "Militants Warn China Over Oil in Niger Delta," *Washington Post* (May 1, 2006).

49. Dino Mahtani, "Nigeria Shifts to China Arms," *The Financial Times* (March 1, 2006).

50. Jon D. Markman, "How China is Winning the Oil Race," *MSN: Money Central* (April 26, 2006). Available at http://moneycentral.msn.com/content/P149330.asp.

51. Richard Akinjide, "Africa, China and Oil and Gas Supplies," *Alexander's Oil and Gas Connections, News and Trends: Africa*, vol. 10, no. 17 (September 15, 2005). Available at www.gasandoil.com/goc/news/nta53701.htm.

52. "China's Trade with Africa Rising Fast," *People's Daily* (November 26, 2004). Available at http://english.people.com.cn/200411/26/eng20041126_165265.html.

53. *Nightly Business Report* (February 22, 2005).Available at www.accessmylibrary.com/coms2/summary_0286-3876750_ITM.

54. Rob Crilly, "Chinese Seek Resources, Profits in Africa," *USA Today* (June 21, 2005). Available at www.usatoday.com/money/world/2005-06-21-africa-china-usat_x.htm.

55. "Background Note: Zimbabwe" (U.S. Department of State official Web site, undated). Available at www.state.gov/r/pa/ei/bgn/5479.htm.

56. "Take Cue from China's Transformation, Zim Urged," *The Herald* [Harare] (May 25, 2005).

57. "Chinese Demand Coal Guarantees," *Financial Gazette* [Harare] (May 20, 2005).

58. "Zimbabwe: Chinese Team Sets Sights On Uranium," *Zimbabwe Independent* [Harare] (May 12, 2006).

59. "Chinese Firms Reportedly Abandoning Projects in Zimbabwe Due to Nonpayment," *Zimbabwe Independent* (May 12, 2006).

60. Paul Nyakazeya, "Government Fails to Cough Up for Planes, Chinese Bitter," *Zimbabwe Independent* (May 16, 2006).

61. See William Kistner, "Timber Trade Fuels Liberia's Misery," *San Jose Mercury News* (August 10, 2003); Allan Thornton, "Chinese Involvement in African Illegal Logging and Timber Trade" (Testimony presented to the House of Representatives Committee on International Relations, before the Subcommittee on Africa, Human Rights and International Operations, June 28, 2005).

62. Craig Timberg, "In Africa, China Trade Brings Growth, Unease," *Washington Post* (June 13, 2006): A14.

63. David Shinn and Joshua Eisenman, "Dueling Priorities for Beijing in the Horn of Africa," *China Brief,* vol. 5, no. 21 (October 13, 2005).

64. Talking Points for Deputy Director-General Qi Jianwei of the Department of West Asia and Africa Affairs of the Chinese Ministry of Commerce at the consultation between the Secretariat of the Chinese Follow-up Committee of the Forum on China-Africa Cooperation (FOCAC) and African Diplomatic Envoys in China (Posted on the China-Africa Cooperation Forum official Web site, May 24, 2005). Available at www.focac.org/eng/zt/zfhzltcsh/t196993.htm.

65. Shinn, "China's Approach to East, North and the Horn of Africa."

66. Nicole Itano, "China's Scale of Cheap Finished Goods Overwhelms African Nations' Struggling Economies," *Newsday* (October 4, 2005). Available at www.newsday.com/business/ny-centerpiece4453961oct04,0,1719138,print.story?coll=ny-business-promo.

67. "Nigerian Body Bans Products from 30 Indian, Chinese, Pakistani Firms," *The Guardian* [Nigeria] (January 7, 2006).

68. "China Reiterates Its Support to COMESA" (Press release: COMESA official Web Site, November 11, 2005). Available at www.comesa.int/news_archive/News_Item.2005-11-11.3013.

69. COMESA was formed in 1994 and includes Angola, Djibouti, Libya, Sudan, Seychelles, Zimbabwe, Kenya, Rwanda, Zambia, Democratic Republic of Congo, Comoros, Ethiopia, Mauritius, Uganda, Burundi, Eritrea, Malawi, Swaziland, Egypt, and Madagascar. The region has a total population of over 350 million and total GDP of over $190 billion. "COMESA Looks Forward to Cementing Ties with China," Xinhua (March 16, 2005). Available at http://news.xinhuanet.com/english/2005-03/16/content_2706628.htm.

70. Ibid.

71. "China to Beef Up Cooperation with Africa's Biggest Economic Bloc," *People's Daily* (November 11, 2005). Available at http://english.people.com.cn/200511/11/eng20051111_220526.html.

72. "A China-SACU FTA: What's In It For SA?" *South African Foreign Policy Monitor* (August–September 2004).

73. Speech by Ambassador Liu Guijin, "China Africa Relations: Equality, Coop-

eration and Mutual Development" (Posted on the Chinese Embassy of South Africa Web site, December 30, 2004). Available at www.chinese-embassy.org.za/eng/znjl/t177587.htm.

74. "China's Trade with South Africa in June 2006," Xinhua (August 11, 2006).

75. "China Pioneers Self-Restrictive Trade with Africa: Expert," Xinhua (June 23, 2006). Available at http://news.xinhuanet.com/english/2006-06/23/content_4740163.htm.

76. "China to Train 10,000 African Personnel in Three Years," Xinhua (December 17, 2003). Available at www.china.org.cn/english/features/China-Africa/82628.htm#.

77. "Focus Is on Aid and Support for Africa," *Business Day* [South Africa] (October 1, 2004).

78. Drew Thompson, "China's Soft Power in Africa: From the Beijing Consensus to Health Diplomacy," *China Brief*, vol. 5, no. 21 (October 13, 2005). Available at www.jamestown.org/publications_details.php?volume_id=408&&issue_id=3491.

79. Ibid.

80. Ibid.

81. "New Agreement Expands Clinical Testing of a Chinese Malaria Vaccine Candidate," PATH Malaria Vaccine Initiative press release (March 15, 2006). Available at www.malariavaccine.org/files/060315-Wanxing-Announcement.htm.

82. "China Signs Tourism MOU with Ethiopia, Mauritius," Xinhua (June 25, 2004). Available at http://news.xinhuanet.com/english/2004-06/25/content_1547764.htm.

83. "Wen Calls for More Help in Africa," *China Daily* (December 16, 2003). Available at http://www1.china.com.cn/english/international/82535.htm#.

84. "S. Africa Sends Largest Tourism Exhibition to China," Xinhua (September 21, 2005). Available at http://en-1.ce.cn/World/Africa/200406/18/t20040618_1100524.shtml.

85. "S. African Diplomat Says 'No Organized Effort' Against Chinese in his Country," Xinhua (February 2, 2006).

86. Liu Guijin speech at the seminar on China-Africa relations held by the Institute for Security Studies of South Africa, "China Africa Relations: Equality, Cooperation and Mutual Development" (November 15, 2004). Available at www.fmprc.gov.cn/zflt/eng/zt/yhjl/t170116.htm.

87. Drew Thompson, "Economic Growth and Soft Power: China's Africa Strategy," *CSIS China Brief* (December 2004).

88. Liu Guijin speech, "China Africa Relations: Equality, Cooperation and Mutual Development."

89. Chris Alden, "Leveraging the Dragon: Toward 'An Africa That Can Say No,'" *Yale Global* (March 1, 2005). Available at http://yaleglobal.yale.edu/display.article?id=5336.

90. "Zim, China Seek Enhanced Economic Ties," *The Herald Online* [Harare] (May 21, 2004).

91. "China Grants Zim Tourism Destination Status," *The Herald Online* [Harare] (December 17, 2003). Examples of Chinese firms' involvement in Zimbabwe include an incident in November 2005 when Harare gave the state-owned China State Farms Agribusiness Corporation farmlands that had been repossessed from white Zimbabweans. Augustine Mukaro, "Chinese Get Farms in Mash West," *The Independent* [Zimbabwe] (November 11, 2005). In 2003, the state-owned China International Water and Electric was awarded a contract to farm 250,000 acres in southern Zimbabwe.

Andrew Meldrum, "Mugabe Hires China to Farm Seized Land," *Guardian Unlimited* (February 13, 2003).

92. "China, South Africa Sign Protocol on Scientific Cooperation," Xinhua (March 13, 2003).

93. "Scaling Up Poverty Reduction Case Studies and Global Learning Process," World Bank Web site. Available at www.worldbank.org/wbi/reducingpoverty/index.html, accessed September 6, 2006.

94. "China Focus: High-level Visits Increase Warmth of China-Africa Relations," *People's Daily* (November 11, 2005). Available at http://english.people.com.cn/200511/11/eng20051111_220446.html.

95. "'State Sabotage' of Radio Station's Broadcasts," Press release: *Reporters Without Borders* (November 10, 2005). Available at www.rsf.org/article.php3?id_article=15561.

96. Liu Guijin speech, "China Africa Relations: Equality, Cooperation and Mutual Development."

97. "More Sports Aids Pledged," *China Daily* (March 21, 2001). Available at www.china.org.cn/e-shenao/olympic/01-05-18.html.

98. Gabriel Rozenberg, Jonathan Clayton, and Gary Duncan, "Thirst for Oil Fuels China's Grand Safari in Africa," *The Times* [London] (July 1, 2006); "China Opens Coffers for Stadium in Guinea," Reuters (May 12, 2006).

99. "Forget Mao, Let's Do Business," *The Economist* (February 7, 2004).

100. "Sino-African Economic, Trade Relations to Be Promoted: Official," *People's Daily* (January 24, 2004). Available at http://english.people.com.cn/200401/24/eng20040124_133246.shtml.

101. "China's Business Links with Africa: A New Scramble," *The Economist* (November 25, 2004). Available at http://economist.com/business/displayStory.cfm?story_id=3436400.

102. "China-African Forum Reaches Action Plan," *China Daily* (December 17, 2003). Available at www.china.org.cn/english/international/82640.htm.

103. Shinn, "China's Approach to East, North and the Horn of Africa" and Ernest Wilson (Testimony before the U.S. House of Representatives Committee on International Relations, Subcommittee on Africa, Global Human Rights and International Operations, July 28, 2005): 56.

104. "More Than Humanitarianism: A Strategic U.S. Approach Toward Africa," *Council on Foreign Relations, Task Force Report No. 56* (January 2006): 43.

105. Shinn, "China's Approach to East, North and the Horn of Africa."

106. Carter Dougherty, "China, Seeking Oil and Foothold, Brings Funds for Africa's Riches," *Boston Globe* (February 22, 2004). Available at www.boston.com/news/world/asia/articles/2004/02/22/china_seeking_oil_and_foothold_brings_funds_for_africas_riches/.

107. Julian Baum and Matt Forney, "Dollar Diplomacy: Beijing, Taipei Woo African States with Cash," *Far Eastern Economic Review* (June 6, 1996): 22.

108. "China Grants 750-Million Dollar Debt Relief to Africa," *People's Daily* (September 6, 2003); "China Provides Aid to Zambia," Xinhua (December 4, 2003).

109. "China Grants 3.72m Dollars in Aid to Niger," Xinhua (December 31, 2005).

110. Craig S. Smith, "World Business Briefing: Asia; China Forgives Some African Debt," *New York Times* (October 12, 2000).

111. "China Grants 750 Million-U.S.-Dollar Debt Relief to Africa," *People's Daily* (September 6, 2003).

112. "Wen Calls for More Help in Africa," *China Daily* (December 16, 2003). Available at www1.china.com.cn/english/international/82535.htm#.

113. Rozenberg, Clayton, and Duncan, "Thirst for Oil Fuels China's Grand Safari in Africa."

114. Zoe Eisenstein, "Angola Sees New Chinese Loans Above $2 Billion," Reuters (November 12, 2005).

115. Rozenberg, Clayton, and Duncan, "Thirst for Oil Fuels China's Grand Safari in Africa."

116. Richard F. Grimmett, "Conventional Arms Transfers to Developing Nations, 1996–2003," *CRS Report to Congress* (August 26, 2004): 39.

117. As of December 2006, Beijing had developed military ties with Sudan, Zimbabwe, Nigeria, Congo, Algeria, Ivory Coast, Ghana, Cameroon, Namibia, Central African Republic, Sierra Leone, Togo, Tunisia, Lesotho, and numerous others.

118. Peter S. Goodman, "China Invests Heavily In Sudan's Oil Industry: Beijing Supplies Arms Used on Villagers," *Washington Post* (December 23, 2004). Available at www.washingtonpost.com/wp-dyn/articles/A21143-2004Dec22.html.

119. "More Than Humanitarianism: A Strategic U.S. Approach Toward Africa," *Council on Foreign Relations, Task Force Report No. 56* (January 2006): 43.

120. In 2005, Beijing sold Sudan 34 new fighter jets; the Sudan air force is equipped with Shenyang fighter planes, including a dozen supersonic F-7 jets. (Reports from *Aviation Week & Space Technology,* cited by Ian Taylor, "Beijing's Arms and Oil Interests in Africa," *China Brief*, vol. 5, no. 21 [October 13, 2005].

121. "China Ready to Step Up Military Cooperation with Sudan," Xinhua (November 28, 2005).

122. "Zimbabwe Buys Military Jets from China," Reuters (April 14, 2005). Available at www.reuters.co.za/locales/c_newsArticle.jsp;:425e75b3:66f57a88bfe3d9?type=topNews&localeKey=en_ZA&storyID=8182075.

123. "Chinese Radar, Anti-Air Missiles for Mugabe's Palace," *New Zimbabwe* (September 9, 2004). Available at www.newzimbabwe.com/pages/palace8.11605.html.

124. "Zimbabwe Reveals China Arms Deal," *BBC News* (June 14, 2004). Available at http://news.bbc.co.uk/1/hi/world/africa/3804629.stm.

125. "Zimbabwe: Editor Discusses State's Purchase of Fighter Jets From China," *Johannesberg Radio 702* [English] (June 10, 2004).

126. Patrick Gilkes, "Arms Pour in for Border War," *BBC News* (March 2, 1999). Available at http://news.bbc.co.uk/1/hi/world/africa/280273.stm.

127. Ethiopian Prime Minister Meles and Chinese Lieutenant General Zhu Wenquan, commander of the Nanjing Military Region, met in Addis Ababa in August 2005. They agreed that "Ethiopia and China shall forge mutual cooperation in military training, exchange of military technologies, and peacekeeping missions, among others." "Prime Minister Wishes to Forge Military Cooperation with China," *Ethiopian Ministry of Foreign Affairs* (August 16, 2005). The previous week Lieutenant General Zhu, joined by Vice Admiral Gu Wengen, met with the commander of the Eritrean Air Force, Major General Teklay Habteselassie. At that gathering, Zhu said that it was China's desire "for the armies of the two sisterly countries to cooperate in various training." "Eritrean Air Force Chief, Chinese Military Delegation Discuss Ties," *Asmara Voice of the Broad Masses of Eritrea* [Eritrean radio] (August 7, 2005). China's training of Eritrean military forces is well known; even Eritrea's President Isaias Afewerki received military training in China. See Frank Ching, "China May Have to Think Again on Africa," *New Straits Times* [Malaysia] (July 14, 2005).

128. John Sorenson, "Lines in the Sand," *Canadian Business and Current Affairs* (June 1, 2005).

129. Daniel Pepper, "Rebels' Arms Made in China," *Washington Times* (April 28, 2006).

130. "Chinese Peacekeeping Troops Leave for Liberia." Posted on Embassy of the People's Republic of China in Australia Web site, dated March 17, 2004. Available at http://au.china-embassy.org/eng/xw/t75244.htm.

131. "Speech by Chinese Premier at Opening Ceremony of China-Africa Cooperation Forum," *People's Daily* (December 16, 2003). Available at http://english.people.com.cn/200312/16/eng20031216_130483.shtml.

132. A sampling of Beijing's African visitors in 2005–6 includes: Mwai Kibaki, President of Kenya (August 2005); Robert Mugabe, President of Zimbabwe (July 2005); Hifikepunye Pohamba, President of Namibia (December 2005); Pakalitha Mosisili, Prime Minister of Lesotho (December 2005); Solomon Berewa, Vice President of Sierra Leone (April 2006); Mohamad Ismail, Deputy General Chief of Staff of the Sudanese Armed Forces (November 2005); Charles Mondjo, Chief of General Staff of the Armed Forces of Congo (December 2005); Kpatcha Gnassingbe, Minister Delegate for the President in charge of Defense of Togo; Aziz Pahad, Deputy Foreign Minister of South Africa (June 2006).

133. "Zimbabwe Report Discussed at UN," BBC (July 27, 2005). Available at http://news.bbc.co.uk/2/hi/africa/4721189.stm.

134. Patrick Goodenough, "Mugabe Envisages Alternative World Order Headed by China," Cybercast News Service [CNS] (December 3, 2003). Available at www.cnsnews.com/ViewForeignBureaus.asp?Page=/ForeignBureaus/archive/200312/FOR20031203a.html.

3

Repaving the Silk Road

China's Emergence in Central Asia

Matthew Oresman

Overview

In October 2002, China held its first joint military exercise in decades with, to the surprise of many, the tiny Central Asian republic of Kyrgyzstan. Aimed at training border forces on both sides to respond to a terrorist insurgency, this event highlighted the growing importance of Central Asia to China.[1] Moreover, this exercise took place with a country that already had United States and Russian forces deployed just outside the Kyrgyz capital of Bishkek. The presence in just one Central Asian nation of Chinese, Russian, and U.S. military and security personnel underscores the convergence of Great Power interests in Central Asia and the renewed strategic importance Central Asia holds for China.

China's engagement in Central Asia has been one of the more intriguing developments of the past decade and a half. China's interest in building relations with Central Asia has a long history dating back to the foundations of the Silk Road. Included in this history are such revolutionary events as the Battle of Talas in 751,[2] and the Chinese conquest of Xinjiang beginning in 1757.[3] The agility and creativity China has exercised in orchestrating its re-emergence in recent years has taken many by surprise, however, especially given the new-found importance of the region to the United States and Russia following September 11, 2001. China has moved rapidly from the difficult task of delineating and disarming its borders with Russia, Kazakhstan, Kyrgyzstan, and Tajikistan, to building a multilateral organization and economic and security ties, all while working to alleviate traditional suspicions among Central Asian states about Beijing's intentions.

Today, China has four principal sets of strategic and diplomatic interests in Central Asia that define its actions there. First, China has strategic and diplomatic goals in Central Asia that play into China's overall foreign policy. Second, China seeks to safeguard the stability and security of borders negotiated after the fall of the Soviet Union. Third, China has focused on preventing

Map No. 3763 Rev.6 UNITED NATIONS
June 2005

Department of Peacekeeping Operations
Cartographic Section

any external influence on affairs in its western region, including support for separatists in Xinjiang. Lastly, China's economic and trade interests in the region, especially the development of energy resources, are also becoming increasingly important to its internal development.

The prominence of China in Central Asia will grow over the next decade, particularly if the Russian position wanes and the strategic attention and deployed capabilities of the United States are drawn elsewhere. On the basis of geography and economic realities alone, China appears well placed to expand its influence in the region over the long run. Central Asian states will continue to seek robust engagement with China as their transportation infrastructure and developing economies become increasingly intertwined. China likely will continue to exercise a light touch with its diplomacy to assure stable, productive relations along its interior frontiers, while dispelling fears that it is seeking regional hegemony.

However, China's current regional position does not yet come close to matching that of the United States, and particularly Russia, in any measurable terms. The United States and Russia provide development assistance, financial aid, and equipment gifts in the billions of dollars. Moreover, both can ably assist the Central Asians in practical responses to terrorist attacks—something China cannot yet do. There are signs that this situation is beginning to shift, however, as China continues to increase its assistance to Central Asia, and the United States attenuates its assistance to some states—particularly Uzbekistan—due to continued human rights violations there and the loss of America's K2 airbase.

Moreover, the United States and China share similar goals in Central Asia, particularly with regard to combating terrorist activity emanating from the region. Beijing recognizes at present that the United States can serve China's interests in Central Asia. Looking ahead, however, Washington, Moscow, and Beijing could find themselves competing for influence in the region as their regional priorities move beyond immediate security concerns to encompass such fundamental issues as Great Power politics, internal political development, and the direction of energy exports. Moreover, China maintains a long-standing concern with "strategic encirclement" by the United States, and Washington remains wary of the implications of China's rise for its global interests. All of these calculations figure into the crafting of China's Central Asian policy and regional engagement plans.

China's Interests and Policies in Central Asia

Strategic Positioning

China seeks to attain several key objectives in its relations with Central Asia. First, at the broadest level, China's approach to Central Asia helps promote its overall diplomatic strategy of establishing a more constructive external environment, while fostering an image of China as a responsible power. The Shanghai Cooperation Organization (SCO), made up of China, Kazakhstan, Kyrgyzstan, Russia, Tajikistan, and Uzbekistan, is a concrete manifestation of this overall foreign policy effort, giving substance to China's widely-touted "new security concept"—updated most recently in the form of China's peaceful rise/development theory—and its emphasis on the Five Principles of Peaceful Coexistence.[4] It provides Beijing an opportunity to demonstrate the value of a multilateral, consultative process versus unilateral or alliance-based approaches to regional security. In establishing and shaping the agenda for the SCO, Beijing has sought to demonstrate its good faith leadership on regional issues of mutual interest.

The SCO is a Chinese-initiated international organization that evolved from the Shanghai Five (named for the five nations sharing the common China-Central Asia frontier) border demarcation and demilitarization process that began after the collapse of the Soviet Union. Initially founded in 2001, the SCO became a full-fledged international organization in 2004 with the launch of a Secretariat in Beijing and a regional counterterrorism center in Tashkent. The SCO plays an integral role in all aspects of China's engagement with Central Asia.

Second, China's relations with Central Asia help Beijing establish stability on its closest periphery so it can focus its attention and assets on internal economic development and other important foreign policy priorities. By and large, Beijing's approach to the region has succeeded in establishing productive political and economic ties that are likely to endure and develop further.

Third, China's policies in Central Asia assist Beijing in managing its bilateral relationships with the other two major powers in the region, Russia and the United States. China seeks to use common interests in Central Asia to strengthen its relationship with Russia, the traditional "big brother" to the region, and foster a strategic environment that matches their respective world views. The SCO represents both the cooperative and competitive nature of the Sino-Russian relationship. The advent of the SCO in 2001 demonstrated Russia's reluctant self-understanding that it could no longer single-handedly maintain Central Asian stability and act as the main external driver of regional developments, and that China had an emerging role to play in the region that Russia could not block. Thus, while the SCO supports shared Chinese and Russian interests, it also acts as a Russian mechanism to monitor and restrain Chinese encroachment into its traditional "backyard." Conversely, the SCO may give China the opportunity to provide a potential alternative to the Russian-dominated Collective Security Treaty Organization (CSTO), which serves as the new collective defense arm for several of the Central Asian states.

China also carries out its policies in Central Asia with an eye on the United States. Beijing's concerns over a growing U.S. presence in Central Asia—beginning with the NATO Partnership for Peace initiatives in the early 1990s; the U.S.-led CENTRASBAT (Central Asian Battalion) peacekeeping exercise in 1997 that transported elements of the 82nd Airborne Division nonstop from Ft. Bragg, North Carolina to the middle of Kazakhstan; and post-September 11 U.S. deployments to Central Asia—underscore its longer-term interests in establishing stronger ties with its Central Asian neighbors to counter a potentially hostile United States presence on China's western doorstep. This concern is directly linked to China's historic fear of strategic encirclement, and has been reflected in various SCO statements, including the 2005 SCO Communiqué calling for an eventual withdrawal of U.S. forces from Central Asia.

Generally speaking, while Russia has always been the most reliable and influential patron of the Central Asian states, leaders know that Russian aid comes with a high price: submission to certain Moscow policies that may not be in the best interest of the individual nations. Likewise, U.S. regional engagement often broaches issues of democratization and human rights, and regional leaders fear that dependence on massive amounts of U.S. assistance may lead to problems should U.S. attention be drawn elsewhere and the flow of aid diminish. China's diplomacy with Central Asia, meanwhile, normally comes without conditions or demands, and thus offers an appealing alternative. By using their partnership with one outside power as leverage against another, Central Asian leaders have managed to increase their strategic options and the amount of assistance they receive from each side.

Regional engagement with China represents a potentially huge boon for both the economic development and security situation of the region. Nonetheless, some Central Asian elites remain worried that China has not revealed its true intentions in the region, and they approach Chinese beneficence with a degree of wariness and caution. Nations such as Kyrgyzstan and Tajikistan are sustained by international assistance, however, and can rarely afford to refuse Chinese offers of aid. Thus, the Central Asian states, while viewing China with prudence, also see the vast opportunity in bringing yet another outside patron into the region.

By and large, Beijing has been successful in leveraging its relationships in Central Asia and within the SCO to achieve its three key strategic objectives. Beijing's interests and policies in Central Asia still face challenges, however. To date, the SCO has been mostly a "talk shop," with few substantive mechanisms to translate words into practice. This may change now that the SCO Secretariat is operational in Beijing and the Regional Antiterrorism Structure (RATS) has begun to function in Tashkent. It is still too early, however, to tell.

Second, China has not entirely given up all of its heavy-handed ways, and has used its size, power, and economic might to gain advantage in negotiations—particularly in discussions over border demarcation and security assistance to monitor the Uyghur diaspora in Central Asia. Central Asians continue to harbor concerns as to China's long-range intentions in the region, which may ultimately limit Beijing's room to maneuver.

Most important, however, is the understanding that while China may offer great potential for economic and security cooperation in Central Asia, the United States and Russia will continue to offer more in the way of concrete security and economic benefits over the near to medium term.

Domestic Security

While China's broad strategic interests provide longer-term direction to its Central Asian policies, specific national security concerns present the most pressing and immediate challenges shaping China's approach to the region. These challenges include the so-called three evils of terrorism, separatism, and extremism, and involve developments within and beyond Chinese borders—from separatist-minded Uyghur groups in China's Xinjiang province, to illicit trans-border activity such as trafficking in drugs, guns, and people. By strengthening its relationships with the Central Asian states, Beijing hopes to combat these problems before they reach the Chinese border. In particular, the SCO provides a prominent platform from which Beijing can comment, and in some cases act, on these security concerns.

China's northwestern province of Xinjiang, known officially as the Xinjiang Uyghur Autonomous Region, presents a unique problem for Beijing.[5] Xinjiang consists of roughly 12 million Uyghurs (approximately 47 percent of the population), a predominantly Muslim Turkic people who were conquered by China in the mid-eighteenth century but not brought under full Chinese dominion until the Communists came to power in 1949. Over the last 50 years, China has exercised a policy of internal colonization manifested in a three-pronged approach: tight political control under the guise of regional autonomy; economic co-option of the Uyghurs by supporting investment, development, and Han migration;[6] and military conquest by the constant presence and application of the People's Liberation Army (PLA), People's Armed Police (PAP), and the paramilitary Xinjiang Production and Construction Corps (XPCC). The Uyghur people have not welcomed Han rule in Xinjiang, and many would like to see either an end to Beijing's control or at least a greater measure of autonomy in their own affairs. While there is no unified Uyghur resistance movement, several small groups do exist, although coordination is poor. Some of these Uyghur groups engage in terrorist activities against the wishes of the numerous nonviolent separatist groups.[7]

Beijing greatly fears the organizing ability of Islamic groups, particularly within the mosques. This fear explains the lengths China has gone to prevent Muslim community groups from interacting with each other and operating beyond the most local level. Islam in Xinjiang, although much less conservative than that practiced in the Middle East, is still considered by Beijing to be the greatest threat to its control of the province. China's nightmare—one actually shared by many Uyghur leaders—is that radical Islamic groups, such as those that moved from the Middle East to Central Asia in the 1990s, will infiltrate Xinjiang and provide aid to the Uyghurs. Of particular significance to China have been reports that al Qaeda and other radical groups have ties to Central Asian Uyghur separatists, such as the East Turkestan Islamic Movement (ETIM). Beijing is also concerned about general support from the Uyghur diaspora living in Central Asia to their beleaguered brothers.

China has sought common cause with Central Asian governments to counter these threats. Some radical groups, such as the Islamic Movement of Uzbekistan (IMU) and Hizb ut-Tahrir (HT), maintain ties to militant Uyghurs in Xinjiang and to pan-Turkic or pan-Islamic radicals elsewhere in Central Asia. China also is interested in ensuring these pan-Turkic or pan-Islamic groups are not able to destabilize the Central Asian states. If these nations fall to radical Islamic leadership, Chinese leaders fear the cross-border effect on China's internal stability. China's primary goal in Central Asia, therefore, is to prevent external conditions from exacerbating an already difficult internal security challenge.

While ETIM operates primarily in Central Asia, the IMU apparently has ties to groups within Xinjiang itself. Between 300 and 700 Uyghurs were detained in Afghanistan during the United States–led war and occupation to oust the Taliban regime following September 11, 2001. Many Uyghurs were found to be present in Afghanistan for training in resistance activities. In fact, these Uyghurs proved to have little affinity for the politics of al Qaeda, but sought more aggressive technical and tactical support for their resistance efforts once they determined that peaceful means of achieving their goals in Xinjiang would prove futile.

The Uyghur diaspora living in Central Asia plays directly into Chinese diplomacy in the region. The diaspora is predominantly concentrated in Kyrgyzstan and Kazakhstan, home to 50,000 and 180,000 Uyghurs, respectively. Many Uyghurs in these countries are entrepreneurs, have achieved middle class status, and are not politically active. Those who are politically active make up a vocal minority. They often petition the government to protect their rights in the face of Chinese pressures to restrict Uyghur activity. Many also provide aid to groups operating within Xinjiang itself, with most of the interaction occurring along very active shuttle trade routes.

The Central Asian states are loath to offend China, and have been proactive in appeasing Chinese worries about the Uyghur populations living in their countries.[8] As one analyst put it, China is having the Central Asian governments do its "dirty work" in the region. For example, on the eve of the first Shanghai Five summit in 1996, the Kyrgyz Justice Ministry prohibited one Uyghur group, Ittipak (Unity), from political activism for three months for failing to curb its "separatist activities," citing the public association provision of the constitution and the noninterference clause of its 1992 communiqué with China.[9] There has been a balancing act between domestic and international pressures, however, as Central Asian regimes do not want to appear to be suppressing a fellow Turkic people.

To stop cross-border cooperation between the Uyghurs in Xinjiang and those in Central Asia, as well as between pan-Turkic and pan-Islamic groups in Central Asia, China promoted the establishment of the SCO Regional Antiterrorism Structure in Tashkent—formally launched in June 2004. It is unclear whether the center will be primarily an information exchange hub like Interpol, or whether it will develop a rapid-response mechanism over time. Since the 1999 Bishkek SCO Summit, when the idea was first proposed, the idea of using the SCO to fight terrorism and other regional security threats has become a centerpiece of the organization and the most salient factor in building practical cooperation to move the organization beyond simply being a discussion forum.

China has also begun to recognize that Central Asia can affect its national security beyond the situation in Xinjiang. Beijing has supported a host of new multilateral and bilateral programs, including a new SCO initiative to address security issues beyond terrorism. Plans are under way to cooperate on emergency response activities, drug trafficking, and law enforcement.[10] The drug initiative will be integrated into the ongoing counterterrorism agenda in the belief that the drug trade is financing terrorist and organized crime groups in the region. Russia, Kazakhstan, and Tajikistan have also discussed the creation of an SCO Rapid Deployment Force (RDF), perhaps mirroring the Russian Collective Security Treaty Organization (CSTO) RDF in Kyrgyzstan. China has yet to support this idea.[11]

Perhaps most significantly, the nations of the SCO (except Uzbekistan) conducted a major, multi-day exercise in early August 2003, simulating responses to various counterterrorism scenarios in Kazakhstan and China. This exercise, held in eastern Kazakhstan and Xinjiang, included over 1,000 troops—many of them special operations forces—and was much more prominent in scope, size, duration, and media coverage than the October 2002 exercise held between Kyrgyzstan and China.[12] More such exercises are planned for the future, possibly in cooperation with the CSTO.

China has initiated a growing bilateral security assistance program with individual Central Asian states, particularly in the areas of border control, military aid, and intelligence sharing.[13] China donated prefabricated border outposts, jeeps, and other monitoring equipment to Kyrgyzstan. China is assisting Tajikistan in guarding their common border as Tajikistan begins to shift resources for the eventual assumption of responsibility for the security of the Tajik-Afghan border from Russian forces. When Uzbek forces faced an IMU flare-up in 2000, China was the first to provide Uzbekistan with emergency military equipment, including flak jackets, night vision equipment, and sniper rifles. China gave 10 million yuan ($1.2 million) worth of military-technical assistance to Kyrgyzstan in 2002, including firearms and telecom systems, to combat terrorism. In February 2003, China donated police facilities to the Internal Affairs Ministry of Kazakhstan. In addition to material aid, China has provided training for Central Asian military officials and continued intelligence sharing—most of which is focused on counterterrorism.

Stability along the Border

A third important set of goals and policies shaping China's active Central Asian diplomacy concerns the settlement of border disputes. Reaching settlements on disputed borders that were sources of tension during the Cold War was important to Beijing in the early 1990s, both to advance its cooperative agenda with Central Asia and to enable China to devote more attention to other post-Cold War strategic challenges. In fact, demarcating and demilitarizing the borders with its Central Asian neighbors (including Russia) became the foundation upon which Sino-Central Asian relations were built. It has provided China and its Central Asian neighbors a measure of peace and security, while setting out a model for cooperative security relations among former adversaries.

The SCO and its predecessor, the Shanghai Five, have played a critical role in legitimizing and institutionalizing these agreements. The most significant accomplishment of the group was the negotiation of the 1996 Shanghai Five Agreement on Confidence-Building in the Military Field Along the Border Areas, and the 1997 Agreement on Reducing Each Other's Military Forces along the Border Regions. These treaties demarcated the border and created a package of military confidence-building measures that included a pullback of troops and equipment to 100 kilometers from common borders, verification procedures, and prenotification requirements for exercises and other military activities near the border.

These agreements also established a series of military-to-military exchanges. The 1996 agreement stipulated that military forces would not be used to attack one another, that military exercises would not be aimed at one another (and would be limited in frequency and scale), and that neighboring Shanghai Five states would be invited to send observers. By the July 2000 Shanghai Five summit, the five parties announced that implementation of the 1996 and 1997 agreements had "helped build for the first time, in the border belt of more than 7,000 km, a region of trust and transparency where military activities are predictable and monitorable."[14]

However, not all border differences have been settled, or settled to all parties' satisfaction. Negotiations continue on a bilateral basis between China and Kazakhstan, Kyrgyzstan, and Tajikistan. China's border negotiations with Kyrgyzstan, for instance, have caused many domestic political problems for the republic. In March 2002, protests erupted in the Asky region of Kyrgyzstan in response to the signing of a border treaty with China. The demonstrators, calling for the resignation of President Akaev, claimed that he had ceded too much to China and had sacrificed Kyrgyz sovereignty. In response to these protests, police tried to quash the demonstration—six people were killed and sixty injured in the melée. This event eventually led to the resignation of the Kyrgyz prime minister and a government investigation. And while the protesters used the charge against China for their own domestic political agenda, the fact that they used a claim against China demonstrates some of the deep-seated resentment toward China held in Central Asian nations. In fact, during the June 2004 Tulip Revolution in Kyrgyzstan, many Chinese shops were looted and burned for no other reason than that they were owned by Chinese.

Still, China has handled its border negotiations with the Central Asians deftly. While China often has received the better bargain, due to its size and power, it has rarely been seen as heavy-handed or offensive, helping to allay fears held by many Central Asian elites that China's true intention is to seek regional hegemony. This is particularly true when examined in light of China's unwavering position when it comes to territorial disputes in the Pacific Ocean. In fact, in the cases of Kyrgyzstan and Tajikistan, China gave into their demands and ceded what had previously been claimed as Chinese territory—a rarity in Chinese history. With border demarcation and demilitarization between China and its Central Asian neighbors virtually complete, remaining border security issues can be placed in the "cooperative security" column of their relationships under the SCO.[15]

Energy and Trade

Finally, China has important economic interests in Central Asia. During his visit to Kazakhstan in 1994, Premier Li Peng called for the construction of a new Silk Road connecting Central Asia with China as a conduit for trade between Asia, the Middle East, and Europe. To date, this plan is still in the developmental phase, although initial upgrades to the transportation infrastructure have already begun.

In 2005, China exported roughly $5 billion worth of goods to Central Asia while importing $3.5 billion from the region.[16] Of the Central Asian states, Kazakhstan has the largest trade arrangement with China, with an approximate total trade value of $7 billion that reflects the natural resources that make up the majority of Sino-Central Asian trade.[17] While these numbers are low in proportion to Chinese trade with others, they are expected to grow as more Kazakh energy resources become available for export. Chinese firms have made some investments in Central Asia, but such financial arrangements have been limited by their risk and complexity, particularly given the heavy-handed presence of gangs and mafia-like extortion rackets that create a hostile investment environment in Central Asia. Chinese investments are wide-ranging—including hotels, factories, and natural resource extraction operations—but the presence of Chinese goods in Central Asian bazaars is ubiquitous (even in Uzbekistan, which is nearly closed to direct trade with China). According to official Kazakh sources, there are now well more than twenty accredited Chinese companies and some 600 joint ventures in the country.[18] While these numbers are low, they are expected to grow, especially as more Kazakh energy resources come online.

Many analysts, therefore, see significant promise in economic and financial relations between China and Central Asia over the medium to long term, especially in the development of the region's enormous energy resources to fuel China's anticipated economic growth and burgeoning energy demands. By 2025, the U.S. Energy Department estimates that Chinese oil demand will nearly double that of today, to 14.2 million barrels a day, with more than two-thirds to be imported.[19] To diversify its sources of supply and increase its energy security, China wants to establish Central Asia—particularly Kazakhstan, and to some degree Turkmenistan—as guaranteed sources of oil and natural gas. In June 1997, China National Petroleum Corporation (CNPC) invested $4.3 billion in the Kazakh state oil company Aktyubinskneft, entitling China to a 63 percent stake in three fields, with a total estimated oil reserve of 1 billion barrels. As part of this agreement, China and Kazakhstan also agreed to build a $3 billion, 3,000-km pipeline from the Caspian Sea area to Xinjiang. The pipeline, when completed, will have an initial output estimated at 10 million tons a year (to double later), and fitting into the East-West Pipeline from Xinjiang's Tarim Basin to China's east coast.

At this point, the pipeline has been deemed uneconomical by many experts, but it reflects China's strategy to secure energy sources at any cost. This is particularly important given the failure of China to secure the import of Russian oil from Siberia. Seeking to diversify its future supply sources, China has continued to invest in Kazakh, Kyrgyz, Turkmen, Uzbek, Tajik, Georgian, and other fields. None of these, however, have offered the same potential as China's main Kazakh field.

As these projects take shape, however, Kazakhstan is exporting small amounts of oil to China. In 2002, China imported nearly 19,600 barrels a day of crude oil from Kazakhstan, representing 1.4 percent of its total imports. This oil, imported primarily by rail, underscores the importance of building more transportation links between China and Central Asia. Such projects include a new rail link being built between Xinjiang and Uzbekistan, which will pass through Kyrgyzstan, and another possible rail link that will include Tajikistan. The intention is to connect this network with the $250 million European Bank for Reconstruction and Development (EBRD)-sponsored Transport Corridor Europe Caucasus Asia (TRACECA) project and the Asian Development Bank's $1.5 billion Central Asia Regional Economical Cooperation (CAREC) transportation, energy, and trade project to erect a new Silk Road linking China to Europe. Additional infrastructure projects include the already completed Urumqi-Almaty rail line and a new 360-km road between Lake Issyk-Kul in Kyrgyzstan and Aksu in Xinjiang—to be built and financed by China.[20]

Central Asia also offers a potential, if smaller, market for China's export-driven economy, particularly as China aims to develop its vast, remote western regions. Chinese goods, while of higher quality and cost than local products, are still much cheaper than Korean, Japanese, Russian, or U.S. items, and thus represent a reasonable purchase for Central Asian consumers. China's development of Xinjiang has in turn provided an import-export destination for Central Asian commodities and raw materials, such as iron ore, steel, and non-ferrous metals; copper to upgrade its power and telecommunication grids; fertilizer for newly irrigated fields; and cattle hides, sheepskins, wools, and cotton fiber. Barter and frontier trade, conducted in large part by Uyghurs, Kyrgyz, and Kazakhs, is also a booming business but is not included in official statistics.[21]

The SCO has begun to play a much more significant role in China-Central Asian economic relations. Since the 2003 St. Petersburg Summit, an ambitious twenty-year plan was launched to reduce barriers to trade and promote regional economic integration. Lead by an SCO working group, a multinational transportation pact is to be completed shortly that should improve the infrastructure of trade in the region. This process has already led to further Chinese investment in road and rail projects, and the construction of new crossings and customs houses along the Chinese border with Central Asia. Four additional working groups have recently been established to address cooperation on e-commerce, customs, quality inspection, and investment promotion.[22] A new $20 million SCO Development Fund and Business Forum was also launched to promote regional trade and investment.

China's new economic diplomacy with Central Asia has led to a sharp increase in the amount of economic assistance, credits, and loans China is providing to Central Asian governments. At the SCO summit held in Tashkent in June of 2004, President Hu Jintao pledged $900 million in credits and loans for Central Asian nations to buy Chinese goods. China provided Uzbekistan with a low-interest loan and a grant of 50 million yuan (over $6 million) for land irrigation projects, and donated $3.62 million worth of medical equipment. China also provided Tajikistan with 25 million yuan (about $3 million) for the implementation of social and economic programs, and Turkmenistan with grants and credits worth 45 million yuan (about $5.4 million) for Turkmengaz to buy equipment and spare parts for its repair plant.[23]

Overall, China's trade with Central Asia has been a boon to the region, and while the amounts are relatively low, the potential for growth is enormous. China's dynamic economy could be a powerful engine for Central Asian development, and its close proximity could provide Central Asian states with an export route to the burgeoning markets of the Pacific. While other nations, such as Japan and Korea, have made headway into the region's markets, it is China's mammoth size and proximity—coupled with Central Asia's natural resource wealth—that will inextricably link China and Central Asia over the coming decades.

The Future of the Shanghai Cooperation Organization

As indicated, the Shanghai Cooperation Organization plays a vital role in all aspects of China's policy toward Central Asia. It has evolved into the main vehicle by which China engages with Central Asia, particularly in the security and counterterrorism area, and recently has become the main driver for regional economic reform (involving China)—particularly in modernizing the economic infrastructure of the region.

China's future successes and failures in Central Asia will be determined, in large part, by the viability of the SCO. Now that the SCO has two working organs—the Secretariat in Beijing and the Regional Antiterrorism Structure in Tashkent—it will be playing an even more active role in regional affairs. This is quite remarkable, given the SCO's humble beginnings as a border negotiation forum, which many experts assumed would fail in its early attempts to evolve into a multifaceted cooperative organization given the robust U.S. presence in the region after September 11, 2001. The SCO has still not had many tangible successes, but the fact that it has endured over the last decade and has just begun a new phase of development bodes well for its future. Yet, member nations will have to commit even more resources, energy, and political capital to make this organization truly viable. The political will seems to exist, but it remains to be seen if Russia and China actually will commit already scarce resources to this effort.

Three early tests will help determine the SCO's future. The first is the Regional Antiterrorism Structure. To be effective, this center will have to be able to coordinate responses to new terrorist threats, with special attention to deconflicting the roles of China, Russia, and the United States. Although the center may not stand up to a new rapid reaction force, it has to be more than an information clearinghouse if the SCO expects to be a respected player on regional security issues.

Secondly, the full establishment and implementation of a permanent Secretariat and budget mechanism will demonstrate the political and material commitment that members are willing to provide the organization. A functioning budget and empowered bureaucracy are central to the success of any international organization, particularly one bringing together such diverse players.

Finally, the SCO must prove that it can accomplish limited economic cooperation, a point stressed with unusual frequency and detail at recent SCO summits. If the SCO can commit to a transportation pact, for instance, it will have proven that it is more than a "talk shop" and can assist in practical economic integration in a troubled region.

The SCO will also need to figure out how it can engage with other nations and multilateral organizations. The 2004 summit laid out a new outline for engaging with partner organizations and granting interested nations observer status. At the 2004 SCO summit, Mongolia became the first SCO observer nation, highlighting Mongolia's growing role as a player in, and partner with, Central Asia. At the July 2005 summit, India, Pakistan, and Iran were added to the group—additions whose long-term effect is unclear, especially given Iran's pariah status in the West, and India and Pakistan's tendency to drown any international organization in which they are members with discussions on Kashmir. President Hamid Karzai of Afghanistan attended both the 2004 and 2005 summits to discuss cooperation with the SCO, including the establishment of an SCO-Afghanistan contact group at the 2005 summit.[24] The emerging presence of these countries in the SCO underscores their growing roles in the development of the security and economic situation of the region. Furthermore, the SCO has signed Memoranda of Understanding with ASEAN and CIS to engage in future cooperation.[25] An agreement with the Collective Security Treaty Organization, most likely in the counterterrorism field, is also probable.[26]

The United States and the SCO

Lastly, how the SCO and the United States construct any future entente could dramatically affect both the future of the SCO and the ability of the Central

Asian nations (and others) to effectively manage the stability and security of the region. There has been talk about the United States becoming an observer in the organization, but this possibility could fundamentally impede the goals of China in the SCO, particularly the use of the organization as a showpiece for Beijing.

This relates to what may be the most striking aspects of the 2005 SCO Communiqué: re-insertion of language opposing "hegemonism" since September 11, 2001, a traditional codeword for the United States; promotion of a "multipolar" world; and a call for the United States to draw up a timetable for giving up its bases in Kyrgyzstan and Uzbekistan once "active" counterterrorism operations end in Afghanistan.[27] Evidence suggests that Uzbekistan initiated the proposal in response to United States and other Western criticism of the government-led massacre at Andijon. The Chinese and Russian governments were content to allow the statement to pass, even though it had no immediate intention to challenge the U.S. presence in Central Asia.[28] A month later, Uzbekistan requested that the United States leave its base at Karshi-Khanabad.

China accepted the statement because it served as a reminder to the United States that China has made impressive gains in Central Asia and has established itself as a less taxing patron than the United States. This message, however, should not be mistaken for an indication that China supports the immediate withdrawal of all U.S. troops from Central Asia or is ready to openly challenge the U.S. presence. China's security interests are still served by the U.S. presence due to its desire for stability in Afghanistan and suppression of radical Islamic movements in Central Asia. China's strategic interests are also not helped by creating tension with the United States over Central Asia when many other issues, such as internal economic growth and the future of Taiwan, are so prominent on the leadership's agenda. This analysis was recently affirmed at the June 2006 SCO summit in Shanghai, where all reference about U.S. bases were removed from the Communiqué and SCO leaders reined in attempts by Iranian President Mahmud Ahmadinejad to use the summit as an opportunity to criticize the United States.

While the departure of U.S. forces from the region may serve the longer-term interests of China (and Russia), SCO members, with the exception of Uzbekistan, are not currently interested in the removal of American bases or going toe-to-toe with the United States in Central Asia. These statements should be recognized for what they are: political and diplomatic signals. SCO member nations are leveraging their combined voice to strengthen their strategic position in relation to the United States to protect their individual and collective interests.

The SCO is not a new Warsaw Pact. It is rather an entity made up of nations that have consistently done what is in their own best interest at the expense of their neighbors for much of the 15 years since independence. Meanwhile, the United States maintains its own relationship with the Central Asian states based on mutual self-interest, a relationship that will not easily be disrupted by the promise of Chinese or Russian patronage.

It is important to remember that the United States achieved more for Central Asian security in the six months following September 11, 2001 than the SCO has done in six years. While there is great cultural affinity for Russia within Central Asia, most do not miss the heavy-handed tactics of their former Soviet ruler. Thus, while Uzbekistan may turn away from the West for fear that the United States and Europe are seeking to undermine the Karimov regime, the new democratic government of Kyrgyzstan, as well as the oil-exporting rulers of Kazakhstan, seem content to engage with the West.

In the end, China's engagement in Central Asia will be heavily dependent on the SCO's development. The goals of the SCO clearly reflect the policy interests of China, and China will continue to use the organization to manage its reemergence as a regional player. Similarly, the Central Asians benefit materially and politically from the organization, and are happy to use Chinese patronage—funneled through the SCO—to their advantage. It is when this internal dynamic shifts away from cooperation to competition, whether it be Sino-Russian tension or an Uzbek-Kazakh schism, that the true strength and endurance ability of the SCO will be revealed.

Bilateral Relationships

While Beijing has established a coherent multilateral strategy in Central Asia, Chinese policy faces different challenges in dealing with each of the countries individually.

Kazakhstan

The China-Kazakh border stretches some 1,533 kilometers, the longest frontier between China and any of the five Central Asian states. China's relationship with Kazakhstan is probably its strongest in the region and best represents China's most basic interests in energy security. Although Kazakhstan and Russia enjoy an extremely close relationship, Kazakh exports to China reached approximately $3 billion in 2004 and represented nearly 75 percent of China's overall trade with the region. Moreover, Chinese President Hu Jintao's June 2003 trip to Kazakhstan—only his third abroad as president—indicated the high priority China places on its relationship with Kazakhstan. This was reaffirmed with the announcement in July 2005 that China and Kazakhstan had formed a strategic partnership.[29] This relationship also includes security and intelligence cooperation, and educational and cultural exchanges, and will continue to grow in the coming years. Much of this growth will be fueled by increased Chinese investment in Kazakh energy and gas, as well as the long-expected construction of a pipeline between the two countries. A recent plan for China to send farmers to rented land in Kazakhstan was not met with the same negative reaction by the Kazakh populace as similar plans in Russia's Far East, which is a significant development. While many Kazakhs are suspicious of Chinese intentions, and disagreements have arisen over China's management of its Kazakh operations, these feelings have rarely affected bilateral relations. Finally, with roughly one million ethnic Kazakhs living in Xinjiang, and 180,000 Uyghurs living in Kazakhstan, Astana takes a particular interest in developments in China's western regions, especially as these two ethnic groups make up a large segment of the shuttle traders.

Kyrgyzstan

With an 858-kilometer border between the two countries, Kyrgyzstan is China's main gateway into Central Asia, and thus represents a point of particular vulnerability for China. Kyrgyzstan has been a source of drugs, organized crime, Islamic radicalism, and Uyghur sympathizers that have had cross-border effects on China's domestic security. From the perspective of the Beijing government, Kyrgyzstan set a particularly disturbing precedent when a populist street movement in May 2005 led to the ousting of President Askar Akaev and establishment of a democratic government in what became known as the Tulip Revolution. The potential implications of such "color revolutions" on the internal affairs of China were not lost on China's leadership. Furthermore, with a U.S.-run base at Manas and Russian-led forces stationed at Kant, Beijing has focused much strategic attention on Kyrgyzstan to ensure the country does not fall outside its influence. As already noted, China's first external military exercise in decades was held with Kyrgyz border forces.

Kyrgyzstan is one of the main transit routes for Chinese goods to the rest of Central Asia—and eventually Europe—and therefore is a key recipient of Chinese infrastructure investment, including new rail and road links. Kyrgyzstan sees China as a potential engine of economic growth and a source of foreign aid. Kyrgyzstan takes its relationship with China very seriously, and, despite some tensions over the Uyghur diaspora and border negotiations, considers China an important partner given its economic and military might. Despite the positive developments in bilateral relations, Kyrgyzstan remains a source of concern for China on issues including drugs, organized crime, Islamic radicalism, Uyghur sympathizers, and perhaps most worrisome in the long run, the presence of U.S. forces just over the border. These concerns came to a head following the Tulip Revolution in May 2005, when President Akaev was deposed and replaced with a democratic government. While Kyrgyzstan's new democratic leaders say they will continue to build strong ties with China, this new political situation may alter the future dynamics of Sino-Kyrgyz relations.[30]

Tajikistan

Having suffered through a long civil war, and still relying heavily on Russian forces to help stabilize its borders, Tajikistan does not yet have the ability to forge a more balanced foreign policy among its various neighbors and allies. Russia remains its primary patron. It is possible that the presence of "gas-and-go" U.S. air operations in Tajikistan signals closer ties to Washington. However, Tajikistan is in dire need of assistance from all quarters, and accepts aid from such disparate countries as the United States, Iran, and China.

With a 434-kilometer border with Tajikistan, China has provided its western neighbor with assistance to secure its border and stem the flow of drugs. Economically, Tajikistan has little to offer China beyond a limited marketplace, but Beijing is keenly interested in making sure Tajikistan's black market—including the massive trade in drugs—does not penetrate into China. While China theoretically has a historic claim to nearly one-third of Tajikistan's territory, China has negotiated the vast majority of border issues in Tajikistan's favor. Overall, Tajikistan's vital ties to Russia dictate much of the relationship and Beijing has failed to solidify strong ties with Dushanbe, but China still maintains active diplomacy there, with an eye on establishing more robust trade ties and preventing Tajikistan from becoming a gateway for drugs and terrorists into China.

Uzbekistan

As the only SCO member not sharing a border with China, its relationship with China has evolved much differently from the others. China has enjoyed decent security ties with Uzbekistan, highlighted by the material aid provided to Uzbekistan during the 2000 IMU attacks in Tashkent and elsewhere. In economic terms, Uzbekistan's harshly protectionist trade policy has blocked many Chinese exports to the region, and the Tashkent government has increased its complaints about shuttle traders bringing cheap Chinese goods into the country.

Uzbekistan's announced goal to become Central Asia's premier power may put it at odds with Chinese interests, particularly its obstinacy to commit more fully to the cooperative security agenda of the SCO. Uzbekistan has consistently played the spoiler in the past for China's engagement with Central Asia. This has begun to change, however, as China has used the majority of the new $900 million Central Asian assistance fund to finance commercial development in Uzbekistan. According to the most recent trade statistics, trade volume increased from almost $60 million in 2001 when Uzbekistan joined the SCO to almost $700 million in 2005—a 1,000 percent increase with a country with massive barriers to trade. While much of this increase represents bank transfers from the new credit line, it is clear that even with protectionist Uzbekistan, China's economic role in the region is increasing.

The full extent of this developing relationship was on display when President Islam Karimov of Uzbekistan made his first trip abroad after the Andijion massacre to Beijing.[31] During the trip, Karimov signed a Friendship Treaty with Chinese President Hu Jintao, and the two announced a $600 million energy deal between China and Uzbekistan. Not only did the visit act as a much-needed show of solidarity for the beleaguered Uzbek president, but it demonstrated to the rest of the developing world that dealing with China does not have the same costs as the United States. Human rights will not be a factor in building relations with China.

Turkmenistan

Turkmenistan has little to offer China on the security front, and export routes for its oil and gas are not conducive to shipment to China. Politically and diplomatically, Turkmenistan has been very difficult to deal with given the closed nature of the government and the bizarre behavior of its leader, the so-called Turkmenbashi, who died in December 2006. Turkmenistan, like Uzbekistan, does not share a border with China, and is even farther from Chinese territory. Its adherence to a policy of "positive neutrality" has kept it out of the SCO. China has invested moderately in Turkmenistan's oil and gas sector, but has limited its ties overall. It appears Beijing will avoid engagement with Ashgabat at least in the near term unless the regional security situation or China's energy needs demand it.

Conclusion: Implications for United States Policy in Central Asia

A key issue for U.S. policymakers in the region—does China's growing prominence in Central Asia negatively affect U.S. interests?—cannot yet be fully answered. At present, and for the near to medium term, the United States should not be overly concerned about China's role in the region, and in some respects should welcome Beijing's approach in Central Asia. There are obvious shared interests between Washington and Beijing in this part of the world, and it is logical for the two, as well as for Moscow, to cooperate in addressing common challenges. These problems include, among others, terrorism, religious extremism, and drug trafficking. The problems of Central Asia are too numerous and too complex to be addressed effectively by the United States alone. Central Asians also lack the capacity to address them individually, and have proven incapable of pulling together more cooperatively to address their political, social, and economic challenges. Washington should encourage multilateral solutions and look for opportunities to engage with China (and Russia) whenever possible. Engaging today will help build the trust and confidence needed to reduce possible tensions over Central Asia that could possibly arise in the future between the three powers, whether it be over the direction of Kazakh energy exports or the remaining U.S. bases in the region.

Nonetheless, the United States should pay close attention to the development of China's relations with Central Asian states to guarantee that China does not become a problematic influence in the region. China's unconditional engagement may serve to undermine U.S. goals of political reform. As Central Asian and Russian oil and gas begin to flow to market, economic competition among the major powers may become more intense, especially because these energy resources will become available during a period in which the United States maintains a regional military presence. Over the long term, should China (and perhaps Iran) become a more dominant presence in Central Asia, the United States will need to remain vigilant to ensure that its interests to promote political development, free trade, economic openness, and stable societies free from extremist influence are not sacrificed in the process.

Overall, China's emergence in Central Asia will continue to grow based on a long history of interaction and apparent mutual interest. China has rediscovered its place in the region and is developing pragmatic channels to achieve its interests there, particularly in the fields of security and natural resource extraction. China's relations with Central Asia in and of themselves, however, are not a major foreign policy priority for Beijing. For the foreseeable future, Beijing's Central Asian diplomacy and strategy will be more a means to other ends; in particular, to promote an alternative multilateral mode of international diplomacy, stable management of Sino-American and Sino-Russian relations, continued domestic economic development, and stability in China's western region—particularly in Xinjiang. By and large, China pays close attention to Central Asia so it does not become a problem. Central Asian nations, meanwhile, have more to gain from the United States and Russia at present than from China, making their ties with China a lower (although growing) priority for them.

China's chief objective is strategic denial—to deny the rise of elements that will challenge China's internal security; deny the use of Central Asia by the United States to contain China; and deny a Russian monopoly of influence on its border. China will most likely give significant attention only to those problems that directly affect its vital interests, such as counterterrorism and other border security and transnational questions. Central to this will be the attention paid to the role of the Uyghurs in China-Central Asian relations.

It is also clear that China's goal is to foster regional cooperation only to the point where it fits into its own national interest. If China can achieve its aims bilaterally, and not through the SCO, it will. The SCO, however, provides a very useful vehicle to address transnational threats and opportunities. Moreover, the region's inability to apply collective action will minimize China's reach, making it likely that Beijing will have considerable influence in some states, such as Kyrgyzstan and Kazakhstan, but less in others.

China's regional position will be most affected by the actions of the United States and Russia. The United States is the most important near-term influence on Central Asian states and thus will have a large impact on the region's security calculations in the future. Russia also has enduring importance to the region, and may undermine many of China's goals if it chooses to exercise its many levers of influence.

In summary, the United States and Russia will have to deal with a more comprehensive Chinese presence in the region in the years ahead. Diplomatic and strategic hedging in the region has already begun. This is not a return to the "Great Game" of the nineteenth century, however. China, Russia, and the United States are too integrated with each other to threaten a clash over what is still a second-tier priority to each of them. However, given the intermingling of Great Power interests in this region, increased tension is possible—and probable. In the best of circumstances, the United States, China, and Russia will recognize their convergence of interests, and work together with the region to shape a more secure, prosperous, and stable Central Asia.

Notes

This chapter is drawn from a larger monograph published in August 2003 by the Freeman Chair in China Studies at the Center for Strategic and International Studies entitled, *China's New Journey to the West: China's Emergence in Central Asia and Implications for U.S. Interests*. It has also appeared in another form as "Beyond the Battle of Talas: China's Re-emergence in Central Asia," in *Tracks of Tamerlane* (Washington, DC: National Defense University Press, 2004).

1. "China Ends War Games with Kyrgyzstan," Associated Press (October 11, 2002); "China, Kyrgyzstan Hold Joint Anti-Terror Military Exercise," Xinhuanet (October 12, 2002); "Joint War Games Boost Terror Fight," *South China Morning Post* (October 12, 2002).

2. In the mid-700s CE, the expansionist Tang spread its influence as far westward as Kabul and Kashmir, eventually coming into direct conflict with the Muslim people of greater Turkestan. At Talas River in 751, a predominantly Muslim army of Arab, Tibetan, and Uyghur forces defeated Chinese troops led by Gao Xianzhi. The ramifications were significant. The Arabs were able to extend their Islamic influence throughout Central Asia and the major trading routes. Tang expansionary tendencies were halted, beginning a trend of military decline. Thus, the Battle of Talas became a demarcation line between the Muslim-Turkic and Chinese worlds, and remains an important touchstone for China-Central Asia relations today.

3. The area known as the Uyghur Kingdom of East Turkestan, created by the migration of Uyghurs from Mongolia and Central Asia during the first millennium, was invaded by the Manchus in 1757, but was not brought under the control of China's Qing (Manchu) Dynasty until 1877 when it was named Xinjiang (meaning "New Territory"). Even then, resistance against Chinese dominion continued until 1949, including the establishment of an independent Uyghur state on two occasions, most notably from 1944 to 1949.

4. China's "New Security Concept" is an evolving foreign policy framework that aims to foster a more equitable, multipolar, and "democratic" international political and economic system worldwide based on informal bilateral strategic partnerships rather than alliances and the Five Principles of Peaceful Coexistence: mutual respect for sovereignty and territorial integrity; mutual nonaggression; mutual noninterference in their respective domestic affairs; mutual benefit; and peaceful coexistence. Recently, this concept has been updated under the peaceful rise/development theory currently being promoted by Chinese leaders. While peaceful rise/development has broader implications, the elements relating to Central Asia—namely a more benign Chinese style of diplomacy aimed at building friendly and lasting ties with neighbors while not making the same demands on the relationship as the United States would (i.e., on human rights and democratization)—are relatively unchanged.

5. For a recent, more detailed description of the Xinjiang issue, see *Xinjiang: China's Muslim Borderland*, ed. S. Frederick Starr (Armonk, NY: M.E. Sharpe, 2004).

6. Termed the "Go West" campaign, Beijing has encouraged the migration of Han Chinese to Xinjiang, and the economic development of its western region, to spread the benefits of China's economic boom beyond its eastern coastline. Aside from the eastern seaboard, in fact, Xinjiang represents the most developed and dynamic economy in the nation. The area has become a center of trade and industry for both China's western region and Central Asia. Xinjiang's greatest potential for growth lies in its natural resources, particularly energy from the Tarim Basin. A 2000 study from the Organisation for Economic Cooperation and Development (OECD) asserts that the Tarim Basin holds at least 3 billion tons of oil in proven reserves and possibly 510 billion cubic meters of natural gas—all thus far untapped. Xinjiang's potential wealth will be critical to developing China's interior and spreading the benefits of economic growth across the country.

7. In response to Beijing's repressive policies in Xinjiang—including several violent crackdowns that, according to some reports, left hundreds dead and thousands imprisoned—and emboldened by the independence achieved by its Central Asian neighbors in the early- to mid-1990s, the Uyghur separatist movement steadily took a more aggressive and violent approach to its goal. On February 27, 1997, three bombs were set off in the Xinjiang capital of Urumqi, killing nine people. Two weeks later,

on March 7, a bomb exploded in Beijing, the seat of Chinese authority, killing thirty. In September 1997, Chinese authorities disclosed that approximately forty small uprisings occurred, with Uyghur activists occupying a half dozen government buildings across China. The forced evictions left eighty dead and two hundred injured. Eight hundred Uyghurs were arrested. Fifteen bomb attacks occurred over a five-month period in 1998 and seven attacks in the first six weeks of 1999, but it is unclear if these were Uyghur actions or just another challenge to the government from some other group—possibly laid-off workers. According to statistics collected by Justin Rudelson, author of the authoritative book on Xinjiang, *Oasis Identities* (New York: Columbia University Press, 1997), there have been more than two hundred militant actions over the past decade resulting in more than one hundred and fifty deaths. These include attacks on police stations, communications, and electric power infrastructure, the bombings of buses, movie theaters, department stores, hotels, markets, and trains, assassinations of judges, and strikes against military bases.

8. Additionally, Uyghur gangs operating in Central Asia, particularly Kyrgyzstan, have been involved in sensational murders and robberies of both Han Chinese and Uyghur peoples—including prominent businessmen and diplomats—keeping this issue center stage for the Chinese and Central Asian authorities.

9. "Temporary Ban on Uyghur Society in Kyrgyzstan," *OMRI Daily Digest* (April 9, 1996).

10. "Tashkent Declaration of the Head of Member States of the Shanghai Cooperation Organization" (June 21, 2004). Available at www.shaps.hawaii.edu/fp/russia/2004/20040621_sco_decl-eng.html.

11. "Tajik President Calls For SCO Rapid-Reaction Force," *Eurasianet's Kyrgyzstan Daily Digest*. Available at www.eurasianet.org/resource/ uzbekistan/hypermail/news/0046.shtml, accessed on June 28, 2005.

12. Malia Dumont, "'Cooperation 2003': Style, Substance, and Some Surprises," *CEF Monthly* (September 2004).

13. While there are no official statistics on Chinese aid to Central Asia, assistance in the form of donated equipment (military and commercial use), grants, and loans has increased dramatically over the last five years.

14. In July 1998, China and Kazakhstan reached a final agreement resolving remaining border disputes along their 1,700-kilometer border—the first full resolution of a border dispute between China and one of its Shanghai Five partners. "China: Jiang Zemin on Nuclear Arms Race, Sino-Kazakh Border Pact," in Foreign Broadcast Information Service, *Daily Report*: China, FBIS-CHI-98-187 (July 6, 1998). The July 2000 quote is drawn from the "Dushanbe Statement of the Heads of State of the Republic of Kazakhstan, the People's Republic of China, the Republic of Kyrgyzstan, the Russian Federation and the Republic of Tajikistan" (July 5, 2000).

15. While all border settlements have been negotiated and approved by the various heads of state, some of the agreements—most notably the Chinese-Tajik border treaty—have yet to be ratified by the legislative branch.

16. *Global Trade Atlas Online*, Global Trade Information Service. Available at www.gtis.com, accessed March 22, 2006.

17. Ibid.

18. "Chinese Business Interests in Central Asia: A Quest for Dominance," *Central Asian-Caucasus Analyst* (June 18, 2003); "Chinese Leader Hu Jintao Visits Astana," *RFE/RL* (June 7 2003); and information provided by background sources.

19. David Sanger, "China's Oil Needs are High on U.S. Agenda," *New York Times*

(April 19, 2004). Available at www.nytimes.com/2006/04/19/world/asia/19china.html.

20. "Chinese President Urges SCO Economic Cooperation," *People's Daily* (May 30, 2003).

21. Ibid.

22. Speech by Premier Wen Jiabao at the Meeting of Prime Ministers of Member States of the Shanghai Cooperation Organization (SCO) (September 23, 2004). Downloaded from Foreign Ministry of the People's Republic of China at www.fmprc.gov.cn/eng/zxxx/t162430.htm on September 28, 2004.

23. Information compiled by the China-Eurasia Forum. Downloaded from www.chinaeurasia.org in October 2004.

24. "Declaration of Heads of Member States of Shanghai Cooperation Organization," (July 5, 2005).

25. "Russia Hails Signing of Document on SCO-CIS Cooperation," Xinhua (April 13, 2005). Available at http://news.xinhuanet.com/english/2005-04/14/content_2825941.htm; "Memorandum of Understanding Between Secretariats of SCO and ASEAN" (April 21, 2005). Available at http://www.aseansec.org/ASEAN-SCO-MOU.pdf.

26. "Putin: CSTO to Establish Anti-Drug Structure," *RIA Novosti*. Available at http://en.rian.ru/world/20050623/40751672.html, accessed on June 23, 2005.

27. "U.S. Urged to Give Bases Deadline," *BBC News Online*. Available at http://news.bbc.co.uk/1/hi/world/asia-pacific/4652175.stm, accessed July 5, 2005; "SCO Sends Strong Signals for West to Leave Central Asia," *People's Daily*. Available at http://english1.people.com.cn//200507/08/eng20050708_194907.html, accessed July 8, 2005; "China-Russia Statement on the 21st Century World Order" (July 2, 2005).

28. "Kyrgyzstani Figures Say SCO Position on U.S. Bases Linked to Uzbekistani Leader," *Interfax–AVN* (July 6, 2005).

29. "Hu Jintao Holds Talks with Nazarbayev and Announces the Establishment of Strategic Partnership between China and Kazakhstan, Chinese Ministry of Foreign Affairs." Available at www.fmprc.gov.cn/eng/topics/hzxcfelseng/t202534.htm, accessed July 7, 2005.

30. For a more complete discussion of this topic, see Matthew Oresman, "Assessing China's Reaction to Kyrgyzstan's Tulip Revolution," *Central Asian-Caucasus Analyst* (April 6, 2005).

31. Chris Buckley, "China Opens Arms to Uzbek Chief," *International Herald Tribune* (May 26, 2005).

4

Hegemony or Partnership

China's Strategy and Diplomacy Toward Latin America

Chung-chian Teng

Introduction

Since the late 1970s, Chinese economic growth has been phenomenal and remains so today.[1] China's rise has ignited debates over the likely effects on international affairs. Scholars and practitioners are closely watching for possible signs of change in China's foreign policy. Will China be a responsible power or threaten world order and peace? Will China challenge the leadership of the United States or cooperate with today's greatest powers? Does the peaceful rise theory advanced by Hu Jintao represent a deviation from Deng Xiaoping's low-profile principles on foreign policy?

There are divergent views about China's rise. Some argue that China has already risen to Great Power status and that it is attempting to "balance or unbalance the predominance of the United States across the globe."[2] Others agree that China has become stronger, but believe that it does not threaten the United States militarily. While acknowledging Chinese military modernization, Colin Powell emphasized the importance of China's integration into the international system and rejected the notion that China's rise inevitably threatens the United States.[3] Henry Kissinger suggested that cooperation between the United States and China would promote peace in the world.[4]

This chapter analyzes China's foreign policy and its behavior in Latin America, which is in the United States' backyard. The following questions are addressed: What is the nature of China's economic interactions with Latin America? What is Beijing's strategy in Latin America? What motivates its efforts to improve relations with regional states? How effective have its policies been, and to what extent are they likely to meet with success in the future?

LATIN AMERICA

803053AI (R02108) 3-04

Table 4.1

China's Merchandise Trade with Latin America, 2001–2005 (unit: $10,000)

	Total trade	Exports	Imports
2001	1,493,890 (18.6%)	823,674 (14.6%)	670,215 (23.9%)
2002	1,782,604 (19.3%)	948,959 (15.2%)	833,645 (24.4%)
2003	2,680,644 (50.4%)	1,187,913 (25.2%)	1,492,730 (79.1%)
2004	4,002,703 (49.3%)	1,824,205 (53.6%)	2,178,498 (45.9%)
2005	5,045,740 (26.1%)	2,368,250 (29.9%)	2,677,490 (23.0%)

Note: Annual growth rates are in parentheses.
Source: Statistics from Ministry of Commerce of the People's Republic of China.

Table 4.2

China's Merchandise World Trade, 2001–2005 (unit: $10,000)

	Total trade	Exports	Imports
2001	50,976,813 (7.5%)	26,615,464 (6.8%)	24,361,349 (8.2%)
2002	62,076,868 (21.8%)	32,556,501 (22.3%)	29,520,307 (21.2%)
2003	85,120,729 (37.1%)	43,837,082 (34.6%)	41,283,647 (39.9%)
2004	115,479,162 (35.7%)	59,336,863 (35.4%)	56,142,299 (42.6%)
2005	142,211,761 (23.2%)	76,199,914 (28.4%)	66,011,847 (17.6%)

Note: Annual growth rates are in parentheses.
Source: Statistics from Ministry of Commerce of the People's Republic of China.

China's Economic Relationship with Latin America

The rapid expansion of the Chinese economy has resulted in growing trade and investment relations with Latin America. This growth is measured against a low base. Even today, China's trade with Latin America represents only a very small share of its total global trade (see Table 4.1 and Table 4.2). China is the world's third largest trading state, but its trade with Latin America was $50 billion in 2005, representing 3.55 percent of its total trade ($1.4 trillion).

Nevertheless, the importance of China-Latin American trade has been growing rapidly. Over the last four years, China-Latin American trade has expanded faster than China's total global trade. In the period of 2001–5, the average annual growth rate of China's trade with Latin America was 32.7 percent, while its global trade grew by 25.1 percent (see Table 4.1 and Table 4.2).[5] The ratio of China-Latin American trade to China-world trade, although still low at 3.55 percent, has increased from 2.93 percent in 2001.

The balance of China's trade with Latin America has shifted dramatically over the last several years, as imports from Latin America have outstripped exports to the region. Chinese imports from Latin America grew 23.9 percent

Table 4.3

China's Top Four Trading Partners in Latin America, 2004–2005
(unit: $10,000)

	Year	Total trade	Exports	Imports
Brazil	2004	1,235,898 (54.8%)	367,485 (71.5%)	868,413 (48.6%)
	2005	1,481,729 (20.0%)	382,755 (31.4%)	998,974 (15.2%)
Mexico	2004	711,270 (43.9%)	497,287 (52.2%)	213,982 (27.6%)
	2005	776,428 (9.2%)	553,827 (11.4%)	222,602 (4%)
Chile	2004	536,480 (51.9%)	168,852 (31.6%)	367,528 (63.5%)
	2005	713,440 (33.2%)	214,989 (27.3%)	498,451 (35.9%)
Argentina	2004	410,723 (29.3%)	85,233 (90.6%)	325,490 (19.3%)
	2005	512,467 (24.8%)	132,509 (55.5%)	379,958 (16.7%)

Note: Annual growth rates are in parentheses.
Source: Statistics from Ministry of Commerce of the People's Republic of China.

in 2001, 24.4 percent in 2002, and 79.1 percent in 2003—significantly faster than the rate of growth of exports to Latin America (at 14.6 percent, 15.2 percent, and 25.2 percent). China began to experience trade deficits with Latin America in 2003.

Within Latin America, China's largest trade partners are Brazil, Mexico, Chile, and Argentina, which together totaled $34.8 billion and accounted for 69 percent of total Sino-Latin American trade in 2005 (see Table 4.3). Because of their economic and political weight in Latin America, China is pursuing free trade agreements (FTAs) with all of them.[6] China's relationship with oil-producing countries in Latin America, most conspicuously Venezuela, is also becoming noticeably closer.

China has also increased its use of foreign direct investment (FDI) in Latin America. Chinese total cumulative investment in this region reached $4.62 billion at the end of 2003, accounting for 14 percent of China's total cumulative investment in the world. In terms of annual FDI flow, China's investment in Latin America in 2004 was $1.7 billion—46 percent of its total.[7] With soaring economic power and intensified economic interaction, the Chinese government must pursue a foreign policy to match and maintain foreign relations capable of supporting its new role in the world.

Rationale Behind Chinese Foreign Relations

Leaders have political and economic rationales for their foreign policies. Politically, Deng Xiaoping's principles remain the most important factors influencing China's policy and behavior. Sustaining continuous development is the economic rationale.

Deng's Peace and Development Perspective

The top priority after Deng's return to power was the Four Modernizations in 1978—a set of developmental objectives that turned a new page in China's contemporary history.

The full-scale policy changes spearheaded at the Twelfth National Congress of the Chinese Communist Party of 1982 included the proclamation of an independent foreign policy. For its external relations, Beijing put some distance between itself and the United States in an effort to take a more balanced position between Washington and Moscow, and reaffirmed its solidarity and cooperation with the Third World.[8]

Gradually, Deng Xiaoping's "peace and development" strategic thinking emerged and was incorporated as a key part of Chinese foreign policy doctrine.[9] While identifying peace and north-south issues as the two major issues in the world in May 1984, Deng proclaimed that the goal of Chinese foreign policy would be the pursuit of world peace for which China would devote itself to modern construction wholeheartedly.[10] Deng did not officially and specifically use the phrase peace and development, however, until June 1985.[11] At the Fourth Plenum of the Sixth National People Congress, held in March 1986, Premier Zhao Ziyang added "peace" into independent foreign policy.[12] China's current foreign policy principles still include the "independent foreign policy of peace," which, in turn, guarantees—according to Chinese foreign policy thought—an international environment conducive to Chinese economic development.[13]

In the aftermath of the 1989 Tiananmen Square incident, Deng elaborated on the future course of the Chinese independent foreign policy by issuing the well-known "28-character guidepost" on several different political occasions.[14] These twenty-eight characters, in fact, represent seven old Chinese sayings: watch and analyze developments calmly; secure our own positions; deal with change with confidence; conceal our capacities; be good at keeping a low profile; never become the leader; and make some contributions.[15] Chinese foreign policy elites have taken these seven Chinese sayings as guiding principles of its foreign policy and guideposts for conducting diplomacy ever since. But their focus has been on "conceal our capacities and make some contributions." Even in 2005, one veteran high-ranking diplomat stressed the lasting importance of "concealing our capacities."[16] In other words, Deng's 28-character guidepost suggested that China adopt a prudent or status quo foreign policy.

New Security Concept and Partnership Under Jiang

Under Jiang Zemin's administration, "peace and development" remained at the center of foreign policy, and Jiang's initiatives focused on establishing partnerships with big powers and the promotion of the "new security concept."

To maintain a long-term stable and peaceful international environment, and improve China's international image, Jiang Zemin promoted the new security concept in 1999. The concept is composed of four core elements, namely mutual trust, mutual benefit, equality, and coordination.[17] General Xiong Guangkai explained, "Mutual trust is the foundation of the new security concept; mutual benefit its objective; equality its guarantee; and coordination the way it operates."[18] This new concept is compatible with the logic of peace and development and new international security norms in the post-Cold War era.

To obtain urgently needed capital and technology, the Jiang Zemin administration pushed for the establishment of partnership relationships with Western powers first. These partnerships are, in Chinese thought, based on equality, mutual respect, friendly cooperation, and lack of confrontation.[19] Chinese decision makers are aware that they must follow the rules in the international market system to gain what they need for further development. They have therefore had to forsake the idea of military partnerships or a kind of military alliance against third parties.[20] China established its new partnerships through various kinds of joint declarations with the major powers, including Russia, France, the United Kingdom, Canada, the European Union, and Japan.[21]

Peaceful Rise Theory Under Hu

Chinese President Hu Jintao first introduced the peaceful rise theory to Asia on his tour of Southeast Asia in October 2003, before his consolidation of political power.[22] Politically, "peaceful rise" implies that "China must seek a peaceful global environment to develop its economy even as it tries to safeguard world peace through development" and will never challenge or replace the world hegemonic powers, nor seek political confrontation.[23] In his 2005 speech in the United States, Zheng Bijian reiterated that China would not pose any threat to other countries or people and would not replace or displace the United States.[24] Furthermore, prior to Hu Jintao's visit to the United States in 2006, Zheng suggested that China would not export its development model to Latin America.[25] Economically, Chinese comments and writing on "peaceful rise" suggest that China will acquire capital, technology, and resources through peaceful means and will not develop its economy through expansion or in a closed way.[26]

However, peaceful rise encourages positive and assertive participation in the international economy. Its supporters argue that China will participate in economic globalization and compete with others in the world market to realize a win-win result, and that China will persist in independent development when participating in economic globalization.[27] Li Junru, vice-president of the Central Party School, stresses the significance of the concept by arguing that China has found an independent development road to build socialism . . .

through involvement in, not isolation from, economic globalization.[28] In addition to their willingness to accept the challenges of economic globalization and competition, the followers of the peaceful rise theory pledge that China will play its dual role as a responsible power.[29]

The peaceful rise theory is the brainchild of President Hu Jintao, who intended to put his own stamp on Chinese foreign policy. But this theory has already been confronted with many challenges from within and without.[30] Some Western observers regard it simply as propaganda aimed at countering the so-called "China threat" theory.[31] Other critics suggest China cannot afford to adopt such a soft approach given the challenges from Taiwan and Japan.[32] Hardliners on the Politburo Standing Committee even proposed an alternative, competing strategy that emerged from a study session in 2004. The study session argued, "To be able to talk peace, one must be able to make war" (*nengyanhe, nengzhanfang*).[33] The basic tenets of this alternative strategy are as follows: (1) a more powerful military will help protect China's interests; (2) China has built increasingly strong foreign and trade relations with neighbors in East, Southeast, and Central Asia, and severely reduced the traditional distrust and suspicions of states in these regions; (3) given these developments, China's military modernization will not cause alarm or spark regional arms races; and (4) civilian industries would potentially benefit from increased investment in military research and development.[34] Owing to its incompatibility with Deng Xiaoping's principles, this alternative national strategy remains under internal discussion.

Peaceful rise remains the core of Chinese current foreign policy, although the term itself has ceased appearing in official statements and documents for now.[35] While visiting the United States in 2005, Zheng Bijian's major theme in his speech at the Asia Society was China's peaceful rise, despite the fact that the Chinese foreign service sector has been very cautious in using the term.[36] Mr. Zheng has a close relationship with President Hu and played a leading role in the development of the peaceful rise theory. Consequently, his speech in New York carries great meaning for understanding Chinese thought.

Facing a rapidly changing economy and society, Hu's foreign policy has to accommodate a challenging domestic and international environment, and he cannot neglect the policy framework set by Deng's "peace and development" strategic thinking and the 28-character guidepost. The Chinese government has placed supreme priority on becoming "a prosperous, democratic, and civilized socialist country." It has established a target GDP per capita of $3,000 by 2050, which would make it "on par with the middle rung of the advanced nations."[37] While such economic goals once required a lower profile foreign policy, the challenges standing in the way of achieving growth under today's circumstances have led Hu to take a more assertive and proactive stand in some areas of foreign policy.

Problem of Economic Scarcity

As the Chinese economy grows, Chinese industries increasingly rely on imported supplies of vital resources. China is now the biggest consumer of copper, tin, zinc, platinum, steel, and iron ore; the second biggest consumer of aluminum and lead; and the third biggest consumer of nickel.[38] The most urgent need is for energy resources. China became a net oil importer in 1993 and the second largest oil consumer and importer in 2004, behind only the United States. China's oil use rose from 4.95 million barrels a day in 2002, to 5.70 million barrels in 2004.[39] In addition, China's cruel oil imports reached 2.6 million barrels a day in 2005.[40] Chinese consumption is projected to reach 12.8 million barrels a day in twenty years.[41] According to the International Energy Agency, China will need to import up to 85 percent of its crude oil and 50 percent of its natural gas by 2030.[42]

With the great expansion of the economy, food supply has also become a high priority. China is the world's top consumer of grain and meat and has even overtaken the United States. In 2004, Chinese consumed 382 million tons of grain and 64 million tons of meat.[43] China's unprecedented economic expansion has forced the Hu Jintao administration to adopt a more assertive foreign economic policy. To guarantee its access to important raw materials and foodstuffs, China has sought to boost its trade and investment relations around the world and, especially, to establish a closer relationship with resource-rich nations.

The Importance of Latin America for China

It is said that the Chinese arrived in America long before Christopher Columbus "discovered" it. According to one Chinese scholar's research, the first arrival dated back to 412 CE when the Buddhist Master Monk Faxian traveled to Mexico after a more than one-hundred day voyage.[44] A more recent study by British naval officer Gavin Menzies asserted that Admiral Hong Bao and Zhou Man, who led two squadrons of Zheng He's armada, reached the coastal area of America around 1421 CE[45] Even if these historical events remain unconfirmed, the interaction between China and Latin America dates back to at least the mid-sixteenth century, when Chinese goods were shipped to Pacific ports in Mexico and Peru by way of Manila in the Philippines.[46]

Political Strategic Consideration

China's partnership with Latin American states is perhaps natural, given that China often identifies itself as a Third World state that has suffered colonial rule and intervention. The Chinese presence in Cuba, as well as in Panama, might illustrate its intention to have some strategic footholds in Latin America. According to a 1999 agreement, China is permitted to operate joint Sino-Cuban signal intelligence and electronic warfare facilities on Bejucal. These facilities are equipped with modern telecommunications and intelligence hardware and are fully integrated into Beijing's global satellite network.[47] A second facility near the city of Santiago de Cuba possesses the capability to intercept, if not necessarily decrypt, classified U.S. military satellite communications.[48] Sino-Cuban military cooperation was strengthened by a cooperation protocol between their armed forces that was signed in December 2000.[49] China shipped arms and explosives to Cuba at least three times over the following year.[50]

Some American political and military elites have expressed concern about the Chinese presence at the Panama Canal. The Panama Ports Company, a subsidiary of Mr. Li Ka-shing's Hutchison-Whampoa Limited (a Hong Kong-based company), signed a contract with the Panamanian government in 1999 for a 25-year lease on the ports at each end of the canal. At the time, U.S. Senate Majority Leader Trent Lott wrote to Defense Secretary William S. Cohen suggesting that China could potentially deny passage to U.S. ships.[51] Former U.S. defense secretary Caspar Weinberger also issued a warning that if China assumed control of this strategic waterway, it would imperil U.S. security.[52]

Turning to a contemporary scenario, a Chinese Navy squadron comprised of the Qingdao missile destroyer and Taicang supply ship made visits to three Latin American countries—Brazil, Ecuador, and Peru—in the Navy's historic first voyage around the world in 2002. The *People's Daily* described the Navy's global circumnavigation as a symbol of the growing overall national strength of China as expressed in the development of the national economy, science, and technology.[53] As a rising power, China demonstrated its willingness and the capability to project its presence across the oceans and around the world.

There is little evidence, however, that present Chinese efforts to promote its relations with Latin America have been mainly based upon their military cooperation or alliances against other nations, except for more frequent military exchanges.[54]

Economic Importance: Raw Materials and Foodstuffs

China looks to Latin America for a steady supply of oil, copper, iron ore, bauxite, and other raw materials, along with soybeans, grains, and other foodstuffs.[55]

It is estimated that Latin America holds 9.7 percent of the world's proven oil reserves and accounted for 13.8 percent of total world output in 2005.[56]

China's Tenth Five-Year Plan (2001–5) suggests that Chinese national oil companies should expand their activities in three strategic regions, Latin America being one of them.[57] Venezuela, which was listed as one of eight target nations for further investment and cooperation in China's Tenth Five-Year Plan, has become one of China's most strategic partners in its search for energy supplies.[58] As China's largest energy supplier in Latin America, Venezuelan oil exports to China grew from 12,000 barrels a day in 2004 to 130,000 barrels a day in 2005.[59] China and Venezuela signed five energy cooperation agreements in early 2005, in which Venezuela promised to provide 100,000 barrels of crude oil per day to China and agreed to a larger role for China National Petroleum Corporation (CNPC) in the development of Venezuela's energy industry.[60]

Latin America is also important as a major source of food, including wheat and beef from Argentina and soybeans from Brazil. Together, Argentina and Brazil produce 42 percent of the world's soybeans and accounted for 45 percent of global soybean exports in 2004.[61] Argentina produced 12.3 million metric tons of wheat and exported 6 million metric tons in 2002–3.[62] China now accounts for 30 percent of Argentine soybean exports.[63] In sum, a third of Chinese agricultural imports were derived from Latin America in 2003, of which Argentina represented 15 percent and Brazil 14 percent.[64]

Implementation of China's Latin America Policy

Following the logic of the peaceful rise theory and the necessity of economic expansion, the focus of China's foreign policy toward Latin America under Hu Jintao has turned to economic development, and the search for oil, raw materials, and foodstuffs has become the dynamic driver.[65]

To cultivate local support for expanding economic relations with the Latin American nations, Chinese leaders have emphasized summit diplomacy. In November 2004, Chinese President Hu Jintao made a tour to four Latin American nations (Argentina, Brazil, Chile, and Cuba), each of which is ruled by a center-left or leftist party and has some ideological affinity with China. President Hu had a formal dialogue with Mexican President Vicente Fox on the sideline of the 12th Economic Leaders' Summit of the Asia-Pacific Economic Cooperation Forum in 2004 and made a state visit to Mexico in the fall of 2005. Vice President Zeng Qinghong made an official visit to the Latin American and Caribbean regions in January 2005.[66] State visits by both the president and vice president within such a short period of time signal the importance of Latin America in Chinese foreign policy.[67]

Recognizing the importance of Latin America's rich resources, China is currently seeking to consolidate its position by emphasizing the win-win nature of its relations with the region. Addressing the National Congress of Brazil during his first Latin American tour in November 2004, President Hu Jintao said relations between China and Latin America should focus on "supporting each other politically and complementing each other economically and building mutual benefits."[68] Later, in Santiago, Mr. Hu indicated that "China, as a large developing country with a strong sense of responsibility, is ready to contribute its share in advancing win-win cooperation for sustainable development."[69] Chinese leaders' rhetoric emphasizes win-win cooperation and non-zero-sum dynamics with Latin America, partly because those principles reflect market economics and partly because of Latin America's strong distaste for anything that smacks of colonialism or neo-colonialism.

Peace and development has served as the core tenet of China's security strategy toward these states. That said, Chinese scholars have also identified other principles as key to China's success during this period: maintaining a peaceful international environment; economic dialogue with the major powers and regional organizations; pursuing multipolarity; and employing economic statecraft.[70]

Partnership in Latin America

Chinese leaders believe that the most important relationships of the twenty-first century will be those between "partners." In Latin America, China has made some strategic moves and followed a kind of Great Power diplomacy to boost its bilateral relationships with the major regional powers. For our purposes, the analysis of Chinese partnerships focuses on its relationships with three nations: Brazil, Mexico, and Cuba.

China-Brazil Partnership

China's close relationship with Brazil is reflected in its cooperation in the areas of trade, resources, technology, and international fora. Trade cooperation between China and Brazil has been growing rapidly over the past several years. Brazil is China's largest trade partner in Latin America and China is Brazil's fourth largest trading partner in the world.[71] Chinese statistics show that China-Brazil trade volume grew 58.4 percent to $9.3 billion between January and September 2004—more than that for all of 2003.[72] President Hu Jintao hopes that annual trade between China and Brazil will grow to $100 billion per year by 2010. The importance of trading with China to Brazil was reflected in Brazilian President Luiz Inacio Lula da Silva's state visit to China in May 2004. Some 421 businessmen, six state governors and seven ministers, one senator and 10 congressmen accompanied the president on the trip.[73]

China's booming economy and pressing need for raw materials have encouraged it to expand its relations with Brazil and create a kind of win-win situation. On the one hand, many Brazilian companies have expanded, or are in the process of expanding, their production capacity to meet China's demand. For example, Brazil's steel industry is investing billions of dollars to increase its capacity by more than 30 percent between 2005 and 2008. On the other hand, Chinese business is planning to make substantial investments in Brazil. For example, the China Minmetals Group is considering a $2 billion investment in Brazil's raw metal material, and Baoshan Steel is joining forces with the Brazilian mining giant Companhia Vale do Rio doce to build a $1.5 billion steel mill in Sao Luis.[74] In the energy sector, China is reaching out to Brazil for its reserves of nonenriched uranium and is interested in convincing Brazilian firms to participate in the construction of eleven new nuclear power plants in China.[75]

In some areas, the relationship goes beyond raw materials. A good example is in aeronautics and space cooperation—especially in earth resource satellites. China and Brazil have deployed two jointly developed earth resource satellites, and they are planning to deepen their cooperation in the international satellite data market. Embraer, Brazil's largest aviation industrial manufacturer, and the China Aviation Industry Corporation II established the joint venture, Harbin Embraer Aircraft Industry Company Limited. The joint venture marks the first time China has cooperated with a foreign civil aircraft manufacturer on an entire aircraft.

China and Brazil also have a common interest in using international fora for political purposes, including efforts to advance the interests of developing states against developed ones. For example, the two worked together closely in the 2003 Cancun World Trade Organization ministerial meeting, playing a shepherding role in the so-called "G-21" to coordinate an "international trade negotiating strategy for the WTO Doha Round."[76] Brazil has also challenged the leadership of the United States in the formation of the Free Trade Agreement of the Americas (FTAA) by strengthening the Common Market of the Southern Cone (MERCOSUR). MERCOSUR has initiated negotiations for signing a free trade agreement with China, and Brazilian President Lula da Silva is soliciting help from China to gain a permanent seat on the United Nations Security Council.

Despite their cordial and close relations, China and Brazil have been entangled in some trade disputes. From 1995–2004, Brazil initiated fifteen antidumping investigations against China.[77] Even after recognition of China as "a market economy" in 2004, the Lula government still decided to look into alleged Chinese unfair trade toward Brazilian industries (including textiles, footwear, toys, and auto parts).[78]

China-Mexico Partnership

China's interest in Mexico lies in gaining access to U.S. markets through Mexico's membership in the 1994 North America Free Trade Agreement (NAFTA), and through Mexico's proximity to the United States.[79] Mexico's location and its network of free trade agreements throughout the region are useful to Beijing in securing access to the rest of Latin America. Mexico is China's second largest trading partner in Latin America and a major investment destination. China has become Mexico's major trading partner in Asia.

The strategic partnership between China and Mexico has been undermined, however, by their competition over the United States market and threats from low Chinese wages and cheap goods to Mexico's economy.[80] It is estimated that from 2001 to 2003, more than 200,000 clothing, textile, and other factory jobs were lost in Mexico to the relocation of factories to China.[81] Leaders in almost every Mexican industrial sector, from traditional handicrafts to assembly-for-export plants, have complained about the penetration of Chinese goods into Mexican markets and jobs lost to low Chinese wages. President Vicente Fox even blamed China publicly for solving its domestic unemployment problem by means of authoritarian and undemocratic policy."[82]

Another potential problem area is human rights. During his presidency, Fox pledged to uncover the abuses during Mexico's dirty war (1960s to 1980s). One of the most sensitive and critical moves was the prosecution of former President Luis Echeverria Alvarez in July 2004 for his connection with the 1968 Tlatelolco Plaza and 1971 Mexico City massacres.[83] Six months later, Chinese Premier Wen Jiabao met Echeverria in Mexico City and praised his contribution toward Sino-Mexican relations, describing him as a friendly ambassador of the people. The visit to Mexico by the Dalai Lama in October 2004 was equally controversial. China was particularly stung by the Dalai Lama's speech before the Mexican Federal Congress, in which he called for the Chinese government to grant total autonomy to Tibet.[84]

However, both countries would like to reverse the downturn in their relationship. The Chinese government, for its part, would like to assure Mexican leaders that the two nations are partners, not rivals. The two countries have established an inter-governmental standing committee to guide and coordinate bilateral cooperation and reduce squabbles over trade and investment.[85] Hu Jintao, as well as Zeng Qinghong, reached agreements with Mexico on enriching and deepening their strategic partnership during their respective visits to the region. During their meeting on the sidelines of the 2004 Asia-Pacific Economic Cooperation forum, Mexican President Vicente Fox endorsed China's bid to join the Inter-American Development Bank and its desire to establish direct air links between Mexico and China.[86]

It will be very difficult to expand their trade substantially and reduce the huge trade surplus (see Table 4.3) enjoyed by China in the short term. The core problem lies in oil transactions and investments. On the one hand, China does not import crude oil from Mexico, because Petroleos Mexicanos (the state-owned oil monopoly) currently cannot increase its crude oil production capacity to meet China's requests.[87] On the other hand, Chinese oil companies cannot acquire the right to explore and produce oil in Mexico through foreign direct investment, because it is prohibited by the Mexican constitution.

China-Cuba Partnership

China's interests in Cuba are more political and ideological than economic. It is the Castro government's need for economic, political, and military support from the outside world that has set the pattern for the relationship. In 1989, Chinese Foreign Minister Qian Qichen and his Cuban counterpart Malmierca exchanged visits and agreed to resume and develop Sino-Cuban relations.

China has provided various kinds of economic support to Cuba. At the fifteenth annual meeting of the China-Cuba Joint Commission on Economic and Trade Cooperation, China Minmetals Corporation signed a long-term procurement contract with the Cuba Nickel Company to buy 2,400 tons of nickel per year. China National Cereals, Oils and Foodstuffs Corporation also signed a contract to buy 400,000 tons of sugar. During Hu's 2004 state visit, China agreed to build a nickel production plant in Cuba and to set up a joint venture to explore Cuban nickel deposits, and to give a 10-year grace period for repayment of loans given to Cuba in the 1990s.[88]

China and Cuba also cooperate on oil exploration and production. In 2005, the China Petroleum and Chemical Corporation (Sinopec) obtained a contract from the Cuban government to explore potentially large oil and gas reserves off Cuba's northwest coast, which could contain as much as 4.5 billion to 9 billion barrels of oil reserves.[89]

For China, despite some growth in the relationship, its trade with Cuba remains negligible. Two-way trade volume reached only $357 million in 2003—some 0.01 percent of China's total two-way trade with Latin America. From China's perspective, military cooperation has been more important—at least from a symbolic point of view.

Dialogue in Regional and Subregional Organizations

Since the 1990s, China's diplomatic activities have increasingly emphasized the importance of multilateralism. Beijing now actively engages with and seeks to develop global, regional, and subregional organizations. According to official Chinese statements, the world order has the following characteristics: the bipolar system is eroding; a multipolar system is emerging; Great Power interaction is intense; and regional and global international organizations play an increasingly important role. China sees the development of a multipolar system as contributing to global peace, stability, and prosperity, and international organizations can contribute to the emergence of such a system.

In Latin America, China has managed to institutionalize its dialogue with the Latin American countries through regional and subregional organizations. In some cases, it has obtained observer status in existing organizations, including the Association for Latin America Integration, the Organization of American States, the Inter-American Development Bank, and the Agency for the Prohibition of Nuclear Weapons in Latin America and the Caribbean. In other cases, it has introduced new political dialogues or consultation mechanisms between itself and existing bodies, such as the Andes Community, The Rio Group, and MERCOSUR. When Vice President Zeng visited Peru in early 2005, the Chinese government arranged a special meeting with the secretary general of the Andes Community as well as the foreign ministers of the member states.[90] It also advanced cooperative agreements such as the Latin America Economic System. Finally, China started to open new dialogue channels. For example, the China-Caribbean Economy and Trade Cooperation Forum was established and held its first ministerial-level meeting in early 2005.

After Tiananmen, Beijing became more involved in UN activities and worked cooperatively within the UN charter system.[91] The Chinese government attempted to foster an image of an independent but responsible member of the community. One important development has been China's increasing participation in UN peacekeeping operations. These activities buttress China's image in the region. In addition, they provide geostrategic benefits, including a growing presence and influence in related regions and in the United Nations.[92] In 2006, China ranked the thirteenth largest contributor of UN peacekeeping forces, with 1,648 troops, police, and military observers stationed in 10 nations.[93]

China has used its United Nations activities to foster better relations with the Latin American states. It not only voted for the UN resolution that sent peacekeeping forces into Haiti in 2004, but also joined the mission by sending 155 police officers to assist and supervise Haitian police, maintain order, and help rebuild the judicial system.[94] China also utilized the opportunity to advance its geostrategic objectives. Chinese Ambassador Wang Guangya visited Haiti in mid-December 2004 to inspect Chinese peacekeeping personnel. But he also used this opportunity to meet with Haitian Interim President Boniface Alexander and Interim Prime Minister Gerald Latortue.[95] Owing to

Figure 4.1 **China's Foreign Exchange Reserves 1999–2006**
(in $ hundred million)

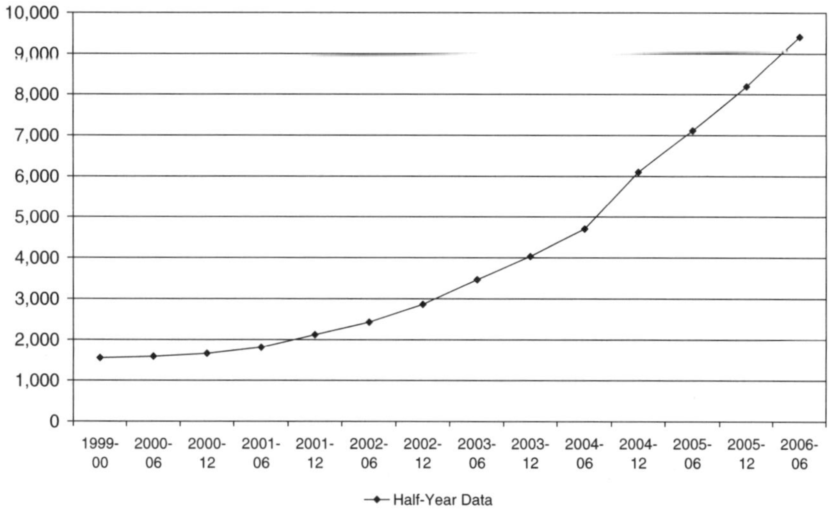

China's key role in the renewal of UN authorization for peacekeeping operations, the Haitian government promised that its prime minister would visit China in the fall of 2005.[96]

Notwithstanding China's active involvement in regional and global multilateral institutions, the United States has taken a cautious and prudent position. For example, the American government sought conditions for China's application for extraregional membership in the Inter-American Development Bank. The main conditions were that "China must make a $200–300 million contribution to the Bank and prepay its International Development Association loans."[97] U.S. action limits the speed of growth of China's presence and influence in the Western hemisphere.

China's Economic Statecraft

The Chinese state has been the single most important factor driving the expansion of economic interactions with the Latin American nations. The Chinese government has wielded its influence directly and indirectly through its cumulative wealth. China's foreign exchange reserves (see Figure 4.1) have soared from almost $155 billion at the end of 1999 to over $941 billion in mid-2006. By providing foreign development assistance, the state has directly intervened in trade and investment transactions. The state has also supplied critical capital to Chinese state enterprises that makes it commercially viable for them to invest in or trade with Latin America.

During the 1980s, Chinese firms focused on setting up joint-venture enterprises—mostly in textiles—in Caribbean countries with behind-the-scenes assistance from the Chinese government. The aim was to establish a local base from which to access or enter the U.S. market.[98] Starting in 1992, when China's Shougang Group invested $120 million in Peru's largest iron ore mine (also the largest in South America), Chinese foreign direct investment began moving into the raw material sector.

In the early twenty-first century, a new type of Chinese investment in Latin America emerged with even more direct intervention by the Chinese government. In a ten-year investment plan announced in November 2004, China pledged to invest more than $19 billion in Argentina, including $8 billion investment in the expansion of Argentina's railway system and $6 billion investment in construction projects.[99] This investment is aimed at facilitating further resource exploitation, production, and transportation, but it is particularly welcome in Argentina, where a 2001 default left the country bereft of major external investors. China made similar commitments to support Brazil's development. In the summer of 2004, Brazilian minister of industry and trade Luiz Fernando Furlan announced that China would invest $5 billion in Brazilian infrastructure projects such as port facilities and railroads.[100] China also pledged to invest new money in Venezuela's energy infrastructure, agricultural development, and telecommunications network (including communication satellites).[101]

China has also sought to guarantee its access to resources in Latin America through strategic alliances, sealed with financial support from the Chinese state. Capital from Chinese state banks has played a crucial role in these relationships. China has been the largest single buyer of Chilean copper, but its position was sealed with a strategic alliance backed by the state.[102] In late May 2005, China's Minmetals Group and the Corporación Nacional del Cobre de Chile (Codelco) agreed to establish a joint-venture company and the maximum investment was to be $2 billion. The agreement awarded a fifteen-year copper supply contract to China and guaranteed financing for future copper projects for Codelco.[103] No less important, Minmetals was awarded the option to buy the shares of the Gaby copper mine.[104] Chile gains stable export revenue and financial loans from China's National Development Bank. China, in return, has the right to develop copper resources.[105]

Aiming to secure stable energy supplies, Chinese investments in Latin American oil have expanded dramatically, mainly through CNPC.[106] CNPC operates three oil fields in Venezuela (i.e., Orimulsion, Caracoles, and Intercampo).[107] CNPC also won the operating rights for petroleum exploration and production at the Talara oil field in Peru from 1993 to 2003.[108] China signed a memorandum of understanding with Peru in early 2005 for new exploration,[109] and signed a Conveyance Agreement with Petroecuador in Ecuador to start exploration and development activities in late 2003.[110]

Chinese companies have also pursued exploration and production deals in Colombia and other Latin American countires.[111] Sinopec and Brazil's Petrobras formed a strategic relationship by signing a comprehensive cooperation agreement on oil exploration, production, refining, marketing, pipes, and technology.[112] They also reached a memorandum of understanding for the construction of a 1225-mile, $1.3 billion natural gas line in Brazil.[113] The Sinochem Corporation acquired a 14 percent share of the Ecuadorian Block 16 oil exploration field.[114]

To gain maximum access to major energy resources, Chinese national oil corporations also cooperate with each other—for example, the formation of the Andes Petroleum Corporation by CNPC, Sinopec, and others. In 2005, Andes Petroleum acquired the concession of five blocks that can produce 75,200 barrels of crude oil per day and have proved reserves of 143 million barrels.[115] This acquisition was made possible through buying the oil and pipeline assets of EnCana Corp, a large Canadian oil-and-natural gas producer.

Notwithstanding China's efforts, it has confronted one major problem in importing oil from South America: the transportation of oil from the Atlantic coast to China.[116] The transportation barriers can be divided into two groups: 1) transportation from the Atlantic coast to the Pacific coast, and 2) the movement of oil from the Pacific coast to China. The first problem is largely one of cost. Gigantic oil tankers cannot pass through the Panama Canal. Under these circumstances, Venezuela is negotiating with Panama to construct a pipeline to the Pacific coast of Panama.[117] The second problem has more to do with security than cost. Beijing worries about the effects of the U.S. naval presence and other maritime factors on the transportation of foreign crude oil back to China.[118] Without any alternative sea routes for South American oil, China must rely on the cooperation and goodwill of the United States to guarantee the security of the Pacific sea lanes, at least in the short term.

China and Taiwan in Latin America

Currently, only twenty-four countries accord Taiwan diplomatic status, and half of them are in Latin America (including the Caribbean nations): Belize, Costa Rica, Dominican Republic, El Salvador, Guatemala, Haiti, Honduras,

Nicaragua, Panama, Paraguay, St. Christopher and Nevis, and St. Vincent and Grenadines. Taiwan pays for this recognition, supplying extensive development aid and disaster assistance.

Ever since 1949, China's external relations have been influenced by its competitive relationship with Taiwan.[119] Chinese scholars agree that the "Taiwan factor" continues to influence China's foreign behavior in general, and Latin America in particular.[120] Some U.S. officials have even suggested that isolating Taiwan is China's key political objective in Latin America.[121] In any case, over the years, China and Taiwan have competed fiercely with each other in Latin America, mobilizing substantial diplomatic resources to achieve their national interests.

With its growing economic power, China gained its foothold in Central America by providing trade and investment opportunities. It established semiofficial Trade Development Offices in Panama, Haiti, and the Dominican Republic, and employed an indirect approach to gain leverage on Taiwan through group pressure. For example, China used an FTA with Brazil and Argentina to motivate these major Latin American states to pressure Paraguay to switch its diplomatic recognition—something they have since done. Panama is the current frontline battleground in that competition.[122] China is now attempting to use its huge commercial potential to influence Panama's diplomatic loyalties, and appears to hope that other Central American countries will follow suit. Although Panamanian officials stress that their government has no plans to recognize the PRC, voices for such a change have been heard from time to time.[123]

Taiwan, for its part, is not entirely passive in defending its interests in the region. In May 2006, Taiwan's President Chen Shuibian visited Paraguay and Costa Rica to shore up support. In Paraguay, President Oscar Duarte joined President Chen in signing a joint communiqué on trade and investment cooperation intended to "promote world peace, democracy, justice and human rights." The communiqué reaffirmed Paraguay's support for Taiwan's bid to join major international organizations, including the United Nations and the World Health Organization.[124] In Costa Rica, Chen attended the inauguration of President-elect Oscar Arias. He then refueled in the Dominican Republic before a surprise visit to Libya. While en route, a defiant President Chen claimed his trip was "for the sake of defending the nation's dignity."[125]

The Effectiveness of China's Policy

To a certain extent, China's Latin American policies have been successful. The various cooperative agreements achieved during the two trips made by

Hu and Zeng demonstrate the extent to which China has improved its position in the region.[126]

In addition to important trade and investment cooperation agreements, China scored three other major achievements. First, Argentina, Brazil, and Chile recognized China's full market economy status.[127] With this recognition, which will make antidumping measures more difficult, analysts believe that their trade and investment interactions will continue growing. Second, China elevated the status of its partnership agreements: with Argentina, the status was elevated from "full-range cooperative" to "strategic"; with Chile and Peru, it pledged to promote the relationship to "full-range cooperative"; and it established a "friendly partnership" with Jamaica. The elevation of partnerships will consolidate their political and economic cooperation. Third, the Chinese government made Brazil, Argentina, Chile, Mexico, Peru, Venezuela, and ten Caribbean nations[128] "Chinese group travelers' destinations."[129] This status can be lucrative for the local economies. According to the World Tourism Organization, China is among the largest sources of tourism now.[130]

Effects on U.S. Interests

The United States remains committed to maintaining firm relations with the nations of the Western hemisphere and would like to help them consolidate both democracy and sustainable economic growth. U.S. President George W. Bush signaled the importance of free trade agreements in the region as a way to further both objectives.[131] With the passage of the Trade Promotion Authority (or "fast track" authorization) by Congress in 2002, the Bush administration started to take positive action to establish free trade areas—particularly those in Latin America.[132]

The free trade network woven by the United States is facing challenges from the major Latin American powers (such as Argentina, Brazil, and Venezuela). These states have divergent views about the future mandate of the Free Trade Area of the Americas, and they differ on a broad range of international questions. Left-leaning governments in Latin America have been very critical of U.S. policy and have sought to form their own community. The South American Community of Nations was officially launched with the signing of the Ayacucho Declaration at the end of 2004, and the Television of the South (Television del Sur, Telesur) began broadcasting Latin American perspectives to the Western hemisphere via satellite in mid-2005.[133]

Eyeing potential capital and business opportunities offered by China, the

Latin American nations have been taking positive actions to promote economic and political intercourse. The bilateral agreements, memoranda, and contracts mentioned earlier in this chapter are important indicators of improving relations between China and the Latin American states. At a minimum, a window of opportunity has opened for China.

The challenges for the United States are great, but will not jeopardize its core national interests in the foreseeable future. As the single largest investing and trading partner with the Latin American nations, the United States has much leverage at its disposal. China's investment remains far behind the total U.S. investment in Latin America.[134] It is fair to say that China is on the rise, but has not yet risen in the region.

Conclusion

With changes in China's domestic and external political economy, the influence of Mao's revolutionary ideology on foreign relations has diminished. The reform and opening policy announced in 1978 was officially approved at the Twelfth National Congress of the Chinese Communist Party in 1982. "Peace and development" was confirmed as the guiding principle in foreign policy and security. Since then, China has been gradually integrated into international society.

With booming economic growth, Chinese decision makers have been more proactive and assertive in formulating foreign policy and conducting diplomacy since the late 1990s. Big power partnerships and the peaceful rise theory reflect the Chinese leaders' strong belief in China's continuing development and its concrete contribution to world peace and order. Implicitly a Great Power, Beijing has the will and capability to assume this role.

Following the logic of "peace and development," China's current strategy toward Latin America appears to revolve around the creation of win-win situations. China uses its rising economic power as a trump card in its diplomacy. Beijing has utilized various kinds of leverage, such as trade, investment, and tourism to improve its regional relations. Two-way trade and investment between China and the Latin American nations have grown dramatically and will likely continue on that trajectory. Chinese businessmen and Chinese tourists are traveling to every part of the region.

The urgent need for various kinds of resources has forced China to shift the focus of its diplomacy to market access for raw materials and energy supplies in particular. In addition to traditional means of increasing trade, Beijing has initiated a series of investment plans to secure market access, including infrastructural construction projects, technological cooperation, and resource exploration.

China and Latin America have complementary economic structures. Vital resources from Latin America will fuel China's booming economy. Inputs from China's purchases and investments will, for their part, facilitate the start of a virtuous cycle of development in Latin America. From a political perspective, the ideological similarity between China and the major powers in Latin America may help China achieve its aims. Leaving nothing to chance, however, China has poured human and material resources into the region.

In view of its rising power, we can expect or even predict that China will continue to pursue policies consistent with the principle of peace and development, and attempt to create or project the image of a good neighbor and a responsible rising power in Latin America. By participating in UN activities, membership in regional international organizations, and formal consultative dialogues in regional fora, Beijing has maintained good relationships and cordial partnerships there. Nevertheless, China should avoid overly aggressive trade practices and consider possible economic losers.

To guarantee its secure access to resources in this region, China may need to increase its military and diplomatic presence. However, China's strategic moves and economic diplomacy will inevitably present challenges to the United States in the Western hemisphere. China's rise has brought changes in the distribution of international economic power, and has enabled it to strengthen its presence in the Western hemisphere. Anti-Washington sentiments have increased in the region and facilitated the rise of Chinese influence and presence. Washington's plans in and for the Western hemisphere may be derailed if it does not increase its attention to the region.

Notes

1. China's Gross Domestic Product (GDP) reached $1.93 trillion in 2004, giving it the world's sixth largest economy. China maintained a 10.3 percent average annual GDP growth rate between 1980 and 1990, 10.6 percent growth from 1990 to 2000, and 9.4 percent growth from 2000 to 2004. One prominent Chinese economist predicted that China "could grow at 9 percent per year for the next 20 years." Between 2001 and 2005, China's exports grew from $266.2 billion to $762 billion and imports from $243.6 billion to $660.1 billion. As of 2004, China had become the world's third largest trading nation, following the United States and Germany. Up to the end of 2003, China's accumulated FDI amounted to $33.4 billion in 139 nations and was the fifth largest source country of FDI. China has also become the world's manufacturing hub. It makes about 30 percent of the globe's television sets and air conditioners, and 50 percent of its cameras.

2. Testimony of Peter T. R. Brookes, Senior Fellow for National Security Affairs and Director of the Asian Studies Center of the Heritage Foundation, before the Subcommittee on the Western Hemisphere, Committee on International Relations, United States House of Representatives (April 6, 2005).

3. "Powell: China Becoming Economic Power," *United Daily News* (June 14, 2005): A13.

4. Henry Kissinger, "China Shifts Centre of Gravity," *The Australian* (June 13, 2005).

5. The growth rates of China-Latin American trade were 18.6 percent, 50.4 percent, and 26.1 percent in 2001, 2003, 2005, respectively. The growth rates of China-world trade were 7.5 percent, 37.1 percent, and 23.2 percent, respectively.

6. Chinese President Hu Jintao and Chilean President Ricardo Lagos Escobar announced the start of talks on a free trade agreement on November 18, 2004.

7. Testimony of Roger F. Noriega, Assistant Secretary of State for Western Hemisphere Affairs, before the Subcommittee on the Western Hemisphere, Committee on International Relations, United States House of Representatives (April 6, 2005). Available at http://usinfo.org/wf/050407/epf406.htm.

8. Wang Jisi, "International Relations Theory and the Study of Chinese Foreign Policy: A Chinese Perspective," in *Chinese Foreign Policy: Theory and Practice,* ed. Thomas W. Robinson and David Shambaugh (New York: Oxford University Press, 1998), 485.

9. Men Honghua, "The International Strategic Framework of China's Peaceful Rise," *World Economics & Politics*, no. 6 (2004): 15; Chen Qimao, "Deng Xiaoping's Diplomatic Thought and Chinese Foreign Policy," in *Chinese Foreign Relations,* ed. Qimao Chen (Taipei: Ji Hong Information, 2000), 38–41.

10. Deng Xiaoping, *Selected Works of Deng Xiaoping,* vol. 3 (Beijing: People's Press, 1993), 56–57.

11. Ibid., 105.

12. Hsu Chih-chia, *Contemporary Chinese Foreign Policy and Sino-American Relations* (Taipei: Sheng-Chih Book, 2004), 65–66.

13. In his closing remark at the first meeting of the Tenth National People's Congress, President Hu Jintao reiterated that China is a peace-loving country and would continue its independent foreign policy of peace. Available at "President Hu Jintao's Talk," *People's Daily* (March 19, 2003): 1.

14. Quansheng Zhao, *Interpreting Chinese Foreign Policy* (New York: Oxford University Press, 1996), 53.

15. Ibid., 53–54.

16. Ambassador Wu Jianmin, the president of China Foreign Affairs University and a veteran diplomat, made this kind of comment on several public occasions. Available at big5.xinhuanet.com/gate/big5/news.xinhuanet.com/world/2005-09/20/content_3516008.htm, accessed on November 29, 2006.

17. Xiong Guangkai, *International Strategy and Revolution in Military Affairs* (Beijing: Tsinghua University Press, 2005), 252.

18. Ibid., 198–200.

19. Joseph Y. S. Cheng and Zhang Wankun, "Patterns and Dynamics of China's International Strategic Behavior," in *Chinese Foreign Policy: Pragmatism and Strategic Behavior,* ed. Suisheng Zhao (Armonk, NY: M. E. Sharpe, 2004), 180–182.

20. Yu Zhengliang et al., *Studies on Great Powers' Strategies* (Beijing: Central Compilation & Translation Press, 1998), 338.

21. Cheng and Zhang, "Patterns and Dynamics of China's International Strategic Behavior," 186.

22. All three most important posts (i.e., General Secretary, President, and Chairman of the Central Military Commission) are in Hu Jintao's hands. Some China experts have indicated that the Hu Jintao era has started. Benjamin Kang Lim, "Hu Replaces Jiang to Cap China's Power Transition," *China Post* (September 20, 2004): 1.

23. Willy Wo-Lap Lam, *Chinese Politics in the Hu Jintao Era* (New York: M.E. Sharpe, 2006), 166.

24. Mainland China News Center, "Zheng Bijian: Beijing not Replace U.S., Don't Regard China as Threat," *United Daily News* (June 19, 2005): A13.

25. "Zheng Bijian: We Export Computers, Rather Than Development Model," *Asia Times Online* (April 11, 2006). Available at www.atchinese.com, accessed on August 10, 2006.

26. Zheng Bijian, "China's Peaceful Rise to Great-Power Status," *Foreign Affairs* vol. 84, no. 5 (Sept./Oct. 2005): 20.

27. Zan Jifang, "Peaceful Rise," *Beijing Review,* vol. 47, no. 16 (April 22, 2004): 19.

28. Ibid.

29. Ibid.

30. Men Honghua, "The International Strategic Framework of China's Peaceful Rise," 19; Chen Qimao, "Deng Xiaoping's Diplomatic Thought and Chinese Foreign Policy," 10.

31. David Shambaugh, "Return to the Middle Kingdom? China and Asia in the Early Twenty-First Century, " in *Power Shift: China and Asia's New Dynamics*, ed. David Shambaugh (Berkeley, CA: University of California Press, 2005), 23.

32. Robert Sutter, "Asia in the Balance: America and China's 'Peaceful Rise,'" *Current History* (September 2004): 286–287.

33. Zhu Feng and Drew Thompson, "New National Strategy Provides Insight Into China's Rise," *China Brief*, vol. 4, no. 17 (September 2, 2004): 3.

34. Ibid., 4.

35. It is widely believed that the foreign service sector prefers to use the term "peaceful development," instead of "peaceful rise."

36. Mainland China News Center, "Zheng Bijian: Beijing not Replace U.S, Don't Regard China as Threat," *United Daily News* (June 19, 2005): A13.

37. Zheng Bijian, "China's Peaceful Rise to Great-Power Status."

38. Robert J. Samuelson, "Great Wall of Unknowns," *Washington Post* (May 26, 2004): A27.

39. Bhushan Bahree, "China Now Leads in New Demand for Oil, IEA Says," *Wall Street Journal* (November 14, 2003): A10.

40. *BP Statistical Review of World Energy 2006* (London: British Petroleum Corporation), 21.

41. Juan Forero, "China's Oil Diplomacy in Latin America," *New York Times* (March 1, 2005): C6.

42. Michael E. Arruda and Ka-Yin Li, "Framework of Policies, Institutions in Place to Enable China to Meet Its Soaring Oil, Gas Demand," *Oil & Gas Journal,* vol. 102, no. 33 (September 6, 2004): 20.

43. Lester R. Brown, "China Replacing the United States as World's Leading Consumer," *Earth Policy Institute* (February 16, 2005).

44. Lian Yunshan, *Who First Came to America: In Memory of Buddhist Master Monk Faxian* (Beijing: Chinese Social Science Press, 1992).

45. Gavin Menzies, *1421: The Year China Discovered America* (New York: Perennial, 2004). Based on the above-mentioned book, the Public Broadcasting System of the United States produced a 120-minute documentary film and broadcast it on July 21, 2004.
46. Yang Jiechi, "China-Latin America Relations in the New Century," Inter-American Development Bank (paper presented at the seminar on The Emergence of China: Opportunities and Challenges for Latin America and the Caribbean, Washington D.C., October 1, 2004).
47. Mary Anastasia O'Grady, "The Middle Kingdom in Latin America," *Asian Wall Street Journal* (September 6, 2004): A7.
48. Ibid.
49. Qi Leyi, "Communist China and Cuba Sign Military Cooperation Agreement," *China Times* (January 2, 2001): 14; "Cuba, China Sign Military Cooperation Protocol," *CNN* (December 28, 2000).
50. Bill Gertz, "China Secretly Shipping Cuba Arms," *Washington Times* (June 12, 2001): A1.
51. Rowan Scarborough, "China Company Grabs Power Over Panama Canal," *Washington Times* (August 12, 1999): 1.
52. George Gedda, "Weinberger Warns China May Try to Take Over Panama Canal," Associated Press (December 7, 1999).
53. Chen Wanjun, Xu Sen, and Mi Jinguo, "Naval Fleet Returns Home from Global Visit," *People's Daily* (September 24, 2002): 4.
54. Jorge I. Dominguez, "China's Relations With Latin America: Shared Gains, Asymmetric Hopes," *Inter-American Dialogue China Working Paper* (June 2006): 6–9; Posture Statement of General Bantz J. Craddock, United States Army Commander, United States Southern Command (Before the 109th Congress House Armed Services Committee, March 14, 2006).
55. Chietigj Bajpaee, "Chinese Energy Strategy in Latin America," *China Brief*, vol. 5, no. 14 (June 21, 2005): 2.
56. *BP Statistical Review of World Energy 2006,* 6–8.
57. Arruda and Li, "Framework of Policies, Institutions in Place to Enable China to Meet Its Soaring Oil, Gas Demand," 20–21.
58. Ibid., 21.
59. William Ratliff, "China and Venezuela: Pragmatism and Ideology" (Testimony before the China Economic and Security Review Commission, August 3, 2006).
60. Bajpaee, "Chinese Energy Strategy in Latin America," 2.
61. Statistics are from the Web site of the American Soybean Association, available at www.systats.com, accessed on July 25, 2005.
62. Statistics are from the Web site of the Australian Wheat Board, available at www.awb.com.au, accessed on July 25, 2005.
63. "China's Soya Needs Lift Argentina," *BBC News* (March 9, 2005).
64. Dominguez, "China's Relations with Latin America," 19.
65. Yang Fuchang and Zhang Qingmin, "The Position and Role of the Developing Countries in the Changing Structure of International Relations and Reflections on China's Policy," *International Studies*, no. 1 (January 2004): 42.
66. Zeng Qinghong led a delegation with one hundred and fifty officials and business representatives and made official visits to Mexico, Peru, Venezuela, Trinidad and Tobago, and Jamaica.
67. After Mr. Hu Jintao assumed the presidency, it has been observed that high-ranking officials reduced their foreign visits to an all-time low.

68. Hu Jintao, "Joining Hands to Enhance Friendship between China and Latin America" (Speech to the Brazilian Parliament, November 12, 2004). Available at www.fmprc.cn, accessed November 14, 2004.

69. "China Vows to Promote Win-Win Cooperation," Xinhuanet (November 19, 2004). Available at www.chinaview.cn, accessed November 20, 2004.

70. Cheng and Zhang, "Patterns and Dynamics of China's International Strategic Behavior," 182.

71. Cynthia Watson, "A Warming Friendship," *China Brief,* vol. 4, no. 12 (June 10, 2004): 1.

72. "Brazil Applauds Market Economy Agreement with China," Xinhuanet (November 13, 2004). Available at www.chinaview.cn, accessed on November 14, 2004.

73. "China and Brazil Ink Joint Communique," *People's Daily* (May 25, 2004): 4.

74. Todd Benson, "China Fuels Brazil's Dream of Being a Steel Power," *New York Times* (May 21, 2004): W1.

75. "China Re-evaluating Energy Policies: Brazil," *China Reform Monitor,* no. 550 (June 23, 2004).

76. Dominguez, "China's Relations with Latin America," 29.

77. Scott Kennedy, "China's Porous Protectionism: The Changing Political Economy of Trade Policy," *Political Science Quarterly,* vol. 120, no. 3 (2005): 413.

78. "Brazil's Industry Probe Can Lead to Possible China Trade Sanctions." Available at http://pravda.ru, accessed on October 20, 2005.

79. Cynthia Watson, "Adios Taiwan, Hola Beijing: Taiwan's Relations with Latin America," *China Brief,* vol. 4, no. 11 (May 27, 2004): 8.

80. James C. McKinley Jr., "Mexico Builds Trade Ties with China," *New York Times* (September 13, 2005): A3.

81. Mary Jordan, "Mexico Now Feels Pinch of Cheap Labor," *Washington Post* (December 3, 2003): A19.

82. Neal Kuo, "Blame Partner for Rising Unemployment: China and Mexico Relations in Troubled Water," *China Times* (August 9, 2004): A10.

83. Kevin Sullivan, "Mexico Prepares to Charge Ex-President," *Washington Post* (July 24, 2004): A1.

84. Neal Kuo, "Dalai Lama Speaks Before Federal Congress: A Test of Sino-Mexico Relations," *China Times* (October 8, 2004): A10.

85. The first conference of the inter-governmental standing committee was convened in Beijing in August 2004 with the signature of two documents—that is, "The Note of the First Conference of the Sino-Mexican Inter-governmental Standing Committee" and the "Sino-Mexican Civil Aviation and Transportation Agreement."

86. "China, Mexico Pledge to Push Forward Strategic Partnership," Xinhuanet, November 21, 2004. Available at www.chinaview.cn, accessed on November 22, 2004.

87. "Mexico Considers Rising Its Oil Output and Exports to China," *People's Daily Online* (February 28, 2004). Available at http://English.peopledaily.com.cn, accessed on August 8, 2006; Petroleos Mexicanos, *PEMEX Statistical Yearbook 2006* (Mexico: PEMEX, 2006), 50.

88. "Cuba, China Sign 16 Agreements," *Voice of America News* (November 23, 2004). Available at www.voa.org, accessed on November 23, 2004.

89. Susan Taylor Martin, "Cuba Seeks Oil Near Keys," *St. Petersburg Times* (May 8, 2006): 1A.

90. This was the highest level meeting between China and the Andes Community after the signing of the Agreement on a Political Consultative and Cooperative Mechanism.

91. Suisheng Zhao, "Beijing's Perception of the International System and Foreign Policy Adjustment after the Tiananmen Incident," in *Chinese Foreign Policy: Pragmatism and Strategic Behavior,* ed. Suisheng Zhao (Armonk, NY: M.E. Sharpe, 2004), 148.

92. Drew Thompson, "Beijing's Participation in UN Peacekeeping Operations," *China Brief*, vol. 5, no. 11 (May 10, 2005): 8.

93. Colum Lynch, "China Filling Void Left by West in U.N. Peacekeeping," *Washington Post* (November 24, 2006): A12.

94. "China Police to Keep Haiti Peace," *BBC News* (September 29, 2004). Available at http://news.bbc.co.uk, accessed on October 12, 2004.

95. Hui Chian Yu and Hsiao-tze Hsu, "Chinese Ambassador Visits Haiti: A Warning Signal to Taiwan's Diplomacy," *China Times* (December 17, 2004): A13.

96. Ibid.

97. Roger F. Noriega, "China Seeks More Influence in Latin America" (Testimony before the Subcommittee on the Western Hemisphere, Committee on International Relations, United States House of Representatives, April 7, 2005).

98. Feng Xu, "China and Latin America After the Cold War's End," in *Latin America in a New World,* ed. Abraham F. Lowenthal and Gregory F. Treverton (Boulder, CO: Westview Press, 1994), 160.

99. "China Plans to Invest $19B in Argentina," Xinhuanet (November 17, 2004). Available at www.chinaview.cn, accessed on November 17, 2004.

100. Peter Howard Wertheim, "Petrobras Signs Strategic Alliance with China's Sinopec," *Oil & Gas Journal*, vol. 102, no. 25 (July 5, 2004): 37.

101. Bajpaee, "Chinese Energy Strategy in Latin America," 2; "Venezuela Promises CNPC More Opportunity to Tap Gas," *Phoenix Television News* (January 31, 2005). Available at www.china5e.com, accessed on July 31, 2005.

102. Yu Mushian, "China, Chile Inks Two Billion Long-term Copper Contract," *Commercial Times* (June 1, 2005): 7.

103. "China's Minmetals and Codelco Sign a Two-billion Cooperative Agreement," Xinhuanet (June 1, 2005). Available at www.xinhuanet.com, accessed on June 2, 2005.

104. Yu, "China, Chile Inks Two Billion Long-term Copper Contract," 7.

105. "China's Minmetals and Codelco Sign Two-billion Cooperative Agreement," Xinhuanet (June 1, 2005). Available at www.xinhuanet.com, accessed on June 2, 2005.

106. South America is one of the three most important areas into which the Chinese petroleum and gas industry has poured a large sum of investment. The other two are the Middle East and North Africa, as well as Central Asia and Russia.

107. China National Petroleum Corporation, *Annual Report 2003,* 20–27.

108. Ibid.

109. Forero, "China's Oil Diplomacy in Latin America," C6.

110. China National Petroleum Corporation, *Annual Report 2003,* 20–27.

111. Forero, "China's Oil Diplomacy in Latin America," C6.

112. Peter Howard Wertheim, "Petrobras Signs Strategic Alliance with China's Sinopec," *Oil & Gas Journal*, vol. 102, no. 25 (July 5, 2004): 35.

113. Peter Howard Wertheim, "Brazil Accepts China as 'Market Economy'; Inks Cooperation Pacts," *Oil & Gas Journal*, vol. 102, no. 44 (November 22, 2004): 33.

114. Sinochem Web site. Available at www.sinochem.com, accessed July 25, 2005.

115. "China's Andes Signs EnCana Deal," *Asia Times Online* (September 17, 2005). Available at www.atimes.com, accessed on August 8, 2006.

116. Bajpaee, "Chinese Energy Strategy in Latin America," 2; Ray Cheung, "Barriers in the Way of Tapping S American Oil and Gas," *South China Morning Post* (November 21, 2004): 7.

117. "Venezuela Planning a Shift of Oil Exports: From U.S. to China," *Phoenix Television News* (February 2, 2005). Available at www.china5e.com, accessed on July 31, 2005.

118. Tarique Niazi, "Gwadar: China's Naval Outpost on the Indian Ocean," *China Brief*, vol. 5, no. 4 (February 15, 2005): 6; Mengdi Gu, "China Wants More Pipelines for Improved Oil Import Security," *Oil & Gas Journal*, vol. 103, no. 1 (January 3, 2005): 59.

119. Peter Van Ness, "China and the Third World: Patterns of Engagement and Indifference," in *China and the World: Chinese Foreign Policy Faces the New Millennium*, ed. Samuel S. Kim (Boulder, CO: Westview, 1998), 160–161.

120. Men Honghua, "The International Strategic Framework of China's Peaceful Rise," 19; Zhang Qingmin, "A Common Policy Toward Unalike Countries: A Preliminary Analysis of China's Policy Toward Developing Countries," *Contemporary China History Studies*, vol. 8, no. 1 (2001): 44–45.

121. Testimony of Peter T. R. Brookes before the Subcommittee on the Western Hemisphere, Committee on International Relations, United States House of Representatives, April 6, 2005. Available at www.heritage.org/Research/TradeandForeignAid/tst040605a.cfm; Testimony of Roger F. Noriega, Assistant Secretary of State for Western Hemisphere Affairs, before the Subcommittee on the Western Hemisphere, Committee on International Relations, United States House of Representatives, April 6, 2005. Available at http://usinfo.org/wf/050407/epf406.htm.

122. Watson, "Adios Taiwan, Hola Beijing: Taiwan's Relations with Latin America," 9.

123. Mainland News Center, " Louder Voices for Diplomatic Relations with PRC among Panamanian Politicians," *United Daily News* (May 16, 2004): A13.

124. "Taiwan, Paraguay Sign Joint Communiqué on Closer Cooperation," Central News Agency [Taiwan] (May 7, 2006).

125. S. J. Tsai and Deborah Kuo, "Taiwan President Blames China 'Political Blackmail' not US for Transit Flap," Central News Agency [Taiwan] (May 7, 2006).

126. President Hu made trips to Argentina, Brazil, Chile, and Cuba as well as participated in the Leaders' Summit of APEC.

127. When joining the World Trade Organization, China agreed that it would be recognized as a nonmarket economy within fifteen years of its entry. Owing to this status, many countries have turned anti-dumping measures into a means of trade protectionism against China; China has become the world's largest anti-dumping target. Consequently, China has made the market economy status issue one of the highest priorities on its economic agenda. "Market Economy Status on Agenda," *China Daily* (April 21, 2004). Available at www.china.org.cn, accessed on November 24, 2004.

128. They are Antigua and Barbuda, Barbados, Bahamas, Guyana, Saint Lucia, Dominica, Suriname, Trinidad and Tobago, Jamaica, and Grenada.

129. This kind of offer has been approved as a very attractive means in China's diplomacy. According to the World Tourism Organization, China is among the top

ten overseas tourist consumption markets and will become the world's fourth largest tourism source nation by 2020. Cuba was approved as a Chinese travelers' destination in 2003. The award of this status will benefit the local economy. "Brazil, Argentina, Chile Approved as Chinese Group Travelers' Destinations," Xinhuanet (November 19, 2004). Available at www.chinaview.cn, accessed on November 19, 2004.

130. "Brazil, Argentina, Chile Approved as Chinese Group Travelers' Destinations," Xinhuanet (November 19, 2004). Available at www.chinaview.cn, accessed on November 19, 2004.

131. President George W. Bush's Opening Speech at the 35th Annual Conference of the Organization of American States General Assembly in Greater Fort Lauderdale, Florida (June 6, 2005). Available at www.oas.org/speeches/speech.asp?sCodigo=05-0113.

132. Up to June 2006, the United States had reached FTAs with the following countries in the Western hemisphere: Canada, Mexico, Chile, Central American nations (Guatemala, Nicaragua, El Salvador, Honduras, and Costa Rica), Dominican Republic, Peru, and Columbia. Several free trade agreements are also under negotiation, including with Panama, Ecuador, and thirty-four Western hemisphere nations as a whole.

133. "Southern American States Become Community," *Youth Daily News* (December 10, 2004): 5; Wang Chiayuan, "Latin America's 'Television of the South' Challenges CNN," *China Times* (July 26, 2005): A11.

134. Roger F. Noriega, "China Seeks More Influence in Latin America" (Testimony before the Subcommittee on the Western Hemisphere, Committee on International Relations, United States House of Representatives, April 7, 2005).

5

China's Interests and Strategy in the Middle East and the Arab World

Mao Yufeng

Introduction

Recent years have seen the People's Republic of China rapidly increase its presence in the Middle East and the Arab world in economic, political, and cultural areas.[1] Economically, negotiations are ongoing with an eye toward establishing a free trade zone between China and the Arab Gulf. The Sino-Arab Cooperation Forum officially opened in Cairo in September 2004, leading to regular meetings of foreign ministers and businessmen.[2] Politically, China appointed a Middle East peace envoy, who has visited the region several times since 2002. Culturally, two China-funded Chinese language schools were established in Egypt and Syria for the first time in history.

This recent activism is motivated by the need to pursue economic resources in the region, particularly energy supplies and markets for Chinese goods and services. This contrasts with China's previous policies in the region, which emphasized political aims at the expense of economic ones.

To assist the channeling of economic resources to the world's fastest growing economy, the Chinese government is pursuing a three-pronged strategy. First, the state functions as an economic actor and uses diplomacy to advance Chinese economic interests in the region. Second, Beijing adopts an offend-no-one policy to regional politics to maintain an environment friendly to Chinese business interests. Third, Beijing exploits its soft power resources to cultivate goodwill and friendship in the region that serves China's long-term economic interests.

China's strategy faces challenges. First, China's goal to be friends with everyone may not always be achievable. Some of China's policies and practices in the region have caused tension in the United States-China relationship. Potential triggers for crises remain and regional conflicts may force Beijing to choose sides between the United States and local actors, or between the local actors themselves. Second, political decisions driven by short-term interests may hurt China's image and, as a result, long-term economic interests in the

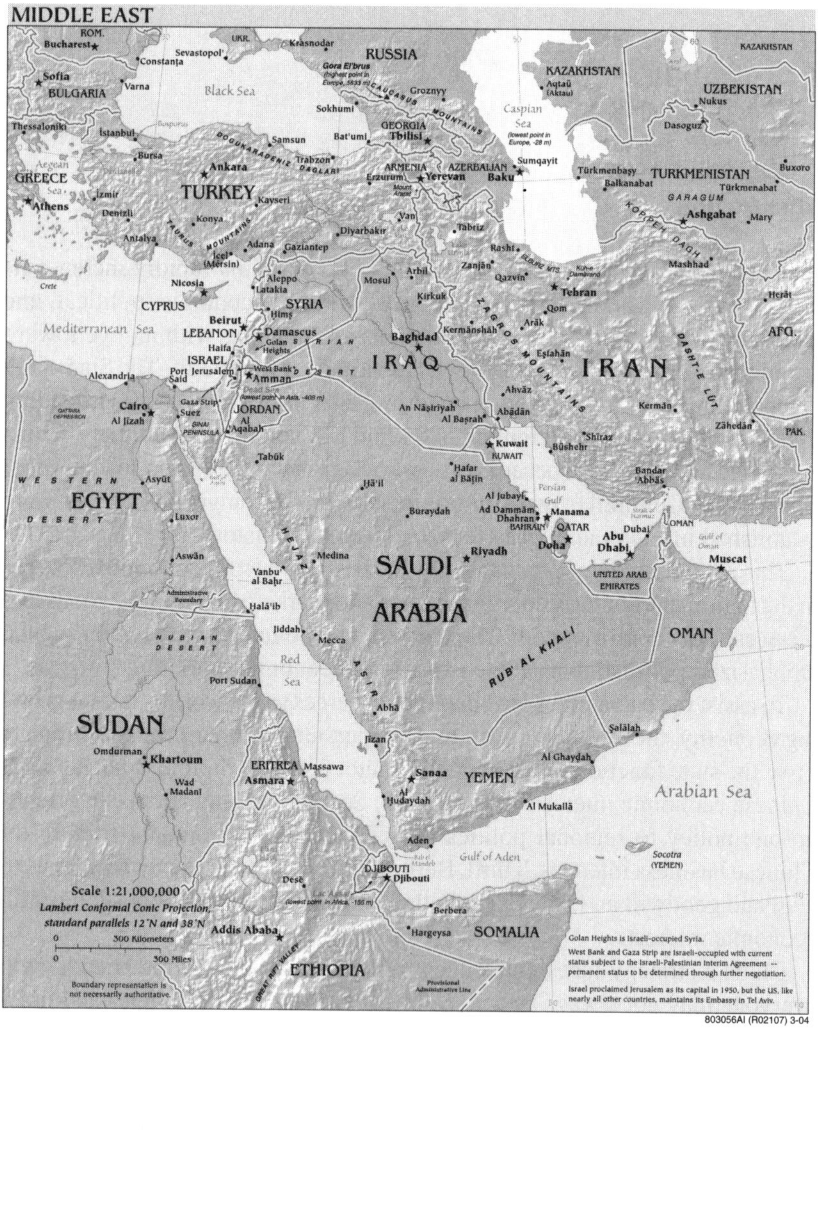

803056AI (R02107) 3-04

region. Third, China's domestic issues, including separatism in Xinjiang, could potentially damage China's relations with countries in the region, thus impeding access to economic resources. All these would pose challenges to Beijing's strategy and result in economic losses.

Historical Context: Profit Is Not the First Consideration

From the founding of the PRC in 1949, until Deng Xiaoping's assent to power in the late 1970s, Beijing's policy in the Middle East and the Arab world was mainly motivated by political goals. Acquiring international recognition, squeezing Taiwan diplomatically, and gaining influence at the expense of the United States and the Soviet Union were China's primary objectives in the Middle East and the Arab world.[3] During this period, China often used trade as a mechanism to advance political goals, and commonly sacrificed economic interests to achieve political ones.

In the 1950s, China's policy in the region was driven by the desire to gain wider recognition for the newly established Communist regime. Before the Bandung Conference of 1955, China pursued a policy of cultivating relations with all governments in the region, including Israel. As a major turning point, China made an important decision at the Bandung Conference to terminate contact with Israel, which was the only country in the region to recognize the People's Republic of China. Realizing the numerical strength of the Arab bloc, China adopted a pro-Arab policy of actively wooing Arab governments including Egypt, Syria, Yemen, Iraq, Morocco, and Sudan. To gain Arab goodwill, China employed various political, economic, and cultural resources, such as foreign aid, cultural and religious exchanges, and support for Arab nationalists in Palestine, Egypt, and Algeria. As one of the first instances of sacrificing economic interests for political gains in the region, China offered to purchase cotton from Egypt in 1953 when it did not actually need imported cotton.[4] Egypt's recognition of the PRC in 1956 touched off a wave of Arab recognition of the PRC in the late 1950s.[5]

Chinese policy took a radical turn after the Sino-Soviet split in the 1960s, when the Soviet Union turned from "Big Brother" to a political competitor in the developing world. China presented itself as a champion of the developing world's struggle against both Soviet and Western imperialism. Moreover, consistent with increasingly radical domestic politics, China's foreign policy in this period was characterized by militant rhetoric and material support for revolutionary groups in Asia and Africa. It became the first non-Arab state to recognize the Palestine Liberation Organization (PLO).[6] Its policy toward Israel became more hostile. During this radical phase, Beijing went so far as to criticize Arab elements that wanted to negotiate with Israel.[7]

Despite extreme economic difficulties at home, China wanted to project an image of itself as a generous aid provider to the region. Although China's donations were small compared to Soviet and U.S. aid packages in the Middle East and the Arab world, China continued to offer aid to Egypt and increased aid to revolutionary groups in Palestine, Algeria, and Yemen.[8]

With the continued decline in Sino-Soviet relations and United States-China rapprochement in the 1970s, China allied itself with American policy to offset Soviet expansion in the region. China readily worked with any anti-Soviet force in the Middle East and moved closer to Western allies such as Lebanon, Iran, and Turkey.[9] At the same time, China toned down its militant rhetoric toward Israel.

Meanwhile, China's long-term effort to alienate Taiwan from the Arab world paid dividends when most Middle Eastern countries voted in favor of the 1971 UN resolution to expel Taiwan and give PRC permanent membership on the Security Council. During this period, Beijing continued to use trade to cultivate ties and goodwill to pave the way for diplomatic relations with countries in the region. Offers made to buy oil from Iraq and Iran in the 1970s, for instance, were largely bait to win the goodwill of the oil-producing countries.[10] Beijing's focus on political goals was best exemplified in Prime Minister Zhou Enlai's comments that "Profit is not the first consideration, but friendship is," during a conversation with the Kuwaiti trade minister in the early 1970s.[11]

Chinese Interests in the Middle East and the Arab World

Because of its huge deposits of energy resources and its potential as a growing market, the Middle East and the Arab world are attracting increased Chinese attention. China now depends on the region for most of its energy imports, and the region has become a large market for Chinese goods and services. This has led to impressive growth in commercial relations with the region and motivates the Chinese government to seek additional ways to attain energy supplies and market share. While political interests remain important, China's recent activism has largely been motivated by economic interests.

Energy

Surging Chinese need for energy has increased China's dependence on oil from the Middle East and the Arab world, which holds two-thirds of the world's oil reserves. Today, roughly half of China's oil imports are from the Middle East.[12] China is making extensive efforts to diversify its energy supplies by looking into sources in Central Asia, Russia, Africa, and the Americas. Chinese analysts predict, however, that despite the region's massive reserves, and the relatively low cost of transportation, oil from the Middle East and the Arab world will continue to constitute the core of China's imports for years to come.[13]

Chinese dependence on oil from this region is so extensive that any supply disruption could have significant negative effects on the Chinese economy. Even speculation on possible disruption has had negative effects. To prepare for the possible disruption caused by the Iraq war, for instance, China increased its purchase of oil in January 2003 by 77.2 percent, in comparison with the previous year, causing the average price to rise by 51 percent. This was an important factor in China's trade deficit in January 2003.[14]

In recent years, China's major oil companies—the China National Petroleum Corporation (CNPC), the China Petroleum and Chemical Corporation (Sinopec), and the China National Offshore Oil Corporation (CNOOC)—have successfully gained some contracts in exploration and production, refining, and infrastructure development in prewar Iraq, Iran,[15] Oman,[16] Algeria, and Libya. For example, Sinopec won a contract in Algeria to invest 75 percent in a $168 million oil exploration and production project in the eastern part of the Sahara Desert in 2002. This was the first time a Chinese company acquired an oil development project in Algeria.[17] In October 2005, CNPC won the first Chinese contract to explore an oil field in Libya,[18] and in September 2005, CNPC's Natural Gas Pipeline department won the contract to build Libya's new pipeline. Worth $34 million, this pipeline will be Libya's largest.[19]

Despite these developments, Chinese companies are late-comers compared with Western oil companies, and are disadvantaged in capital and technology. As such, they own relatively few exploration and production assets in the region. Chinese analysts believe that Chinese oil companies need to expand their involvement in exploration and production activities to reduce the cost of purchasing oil, and that this is an area requiring government support.[20]

Markets

In addition to being a significant source of China's energy security, the Middle East and Arab world are important markets for Chinese products and services. During the past decade, Chinese exports made significant headway in several countries in the region. The Arab market still has great potential for growth and the Chinese are eyeing greater market share and a higher margin of profit.

Chinese exports to the region increased by 500 percent in the past decade. In addition to the growth of absolute numbers, Chinese products have gained increasingly larger market shares. For example, Chinese products constituted 37 percent of the total imports of Dubai, UAE in 2004.[21] This was twice the market share of South Korea, which was the second largest exporter to Dubai.

Most of China's exports to the Middle East and the Arab world are to the Persian Gulf. In 2005, the trade value between China and Saudi Arabia, Kuwait, Bahrain, Qatar, the United Arab Emirates, Oman, and Iran reached $32 billion, which was approximately 70 percent of the total $46 billion trade value between China and the entire Middle East and the Arab world.[22] Among these countries, the UAE is currently the largest destination for China's exports to the region, valued at $6.8 billion in 2004.[23] In 2004, Iran supplied 15 percent of China's oil imports and was China's largest overseas market for construction projects, with over 120 large-scale projects underway in all parts of the country.[24] Because of this, Beijing's commercial interests to date have largely been focused in the Gulf region.

Despite the impressive growth of Chinese exports to the Persian Gulf region, China's market share in the greater Middle East is still relatively small. For example, Chinese exports to the Arab world account for less than 4 percent of the Arab countries' total imports. In addition, Chinese exports to the region mainly consist of lower- and intermediate-end consumer products, including machinery and electronic products, clothes, textiles, shoes, tires, and bags.[25] Because the profit margin for such products is low, this has caused harsh competition among Chinese firms, generally resulting in loss of profit and damage to the image of Chinese products in the region. Chinese analysts call for the Chinese government to regulate competition and facilitate coordination among Chinese companies in the region. Moreover, in order to secure greater market share in the region, the Chinese government is using incentives to encourage the export of higher-end products such as computers and automobiles.

Increasing imports of oil and export of manufactured goods and services have rapidly increased overall trade volumes with the region. In 2004, the trade volume between China and the twenty-two Arab countries reached $36.7 billion[26]—about the same as the trade volume between the United States and the Arab world. In the past few years, this number has been growing at an annual rate of over 40 percent,[27] and 2005 witnessed yet another upsurge in trade with the region. In Saudi Arabia alone, trade during this year increased by 56 percent, with a 62.8 percent increase in imports.[28]

Political Interests

For its numerical strength, the region continues to be regarded as vital to China's political interests. The Chinese government counts on the region to continue supporting the one-China principle in relation to Taiwan.[29] Moreover, positive political relations with governments in the region help generate favorable votes in international organizations such as the United Nations and the World Trade Organization. At the annual Geneva meeting of the UN Human Rights Commission, for example, countries in the region consistently sided with the PRC on the issue of human rights, effectively shielding Beijing from attempts to vote on a censure motion that would criticize the regime for abuses. Barring dramatic changes in regional politics, the Chinese government expects friendly political relations with regional governments to continue. As such, interest in the economic resources of the region has motivated the Chinese government's recent activism there.

China's Strategy in the Middle East and the Arab World

Beijing has adopted a three-pronged strategy to promote Chinese interests in the Middle East and the Arab world. First, the Chinese state uses diplomacy to help Chinese businesses make inroads to the oil industry and markets of the Middle East and the Arab world. Second, Beijing adopts an offend-no-one approach to regional politics in order to maintain an environment friendly to Chinese business interests. Third, Beijing exploits its soft power resources to cultivate goodwill and friendship with an eye toward protecting China's long-term economic interests. This three-pronged strategy reflects China's foreign policy priorities and its efficient use of limited resources.

Economic Diplomacy

China's increasing trade volume with the region is partly the result of a state-led effort to promote economic relations with the region. Operating as an economic actor, the Chinese state promotes Chinese economic interests through an increasingly engaged apparatus of government, personal, and semi-governmental institutions. The functions of this state apparatus are twofold. At home, it encourages and facilitates trade through financial incentives and services, managing the trade imbalance[30] and offering research assistance.[31] In the region, diplomacy is used to secure Chinese economic interests. This includes general diplomatic efforts to promote trade, a focused effort in the Persian Gulf region, and special relations with countries considered "problematic" by the West.

China has reached trade and investment protection agreements with all Arab and Middle Eastern governments, with the exception of Somalia. In the 1980s, China established the Joint Commission on Commerce and Trade.The commission brings together government officials, diplomats, business representatives, and experts from China and the region to improve understanding and communication. The Joint Sino-Arab Chamber of Commerce, established in 1988, has branches in all twenty-two Arab countries. Government-sponsored business associations in China have mushroomed and are engaged in activities that include hosting Middle Eastern trade exhibitions and organizing Chinese economic delegations to visit the Middle East and the Arab world.

Beijing's recent diplomatic overtures have made the Persian Gulf a focal point as Beijing pushes for higher levels of cooperation and greater integration with the Arab economies. In addition to signing the "Framework" with Iran,[32] Beijing is negotiating with the Gulf Cooperation Council to establish a free-trade zone. During Saudi King Abdullah bin Abdul-Aziz's visit to Beijing in January 2006, Chinese Premier Wen Jiabao called for "greater two-way investment in energy, trade, and telecommunication sectors."[33] During Abdul-Aziz's visit, the Chinese and Saudi governments also signed five agreements regarding cooperation in oil, gas, mineral deposits, trade, loans, and professional training.[34]

Chinese diplomats in the Gulf region have made advancing business interests a priority among their responsibilities. For example, the first Chinese ambassador to Iraq after the fall of Saddam Hussein's regime indicated that one of his main goals was to help Chinese companies gain reconstruction contracts.[35] Not only are they involved in deal-making, but Chinese diplomats often act as spokespeople for Chinese businesses.[36] Moreover, diplomats and businessmen form an extensive network which functions as an apparatus for promoting Chinese economic interests in the region.[37] Hoping to enjoy continuing "guidance and support"[38] from the government, Chinese businesses actively engage diplomats in their activities. For example, the Association for Chinese-invested Enterprises in Iran has made the Chinese commercial attaché its honorary leader,[39] and the head of the China-Saudi Friendship Association, established in 1997, is China's former oil minister, Wang Tao.

Beijing's diplomacy in the Gulf region seeks diversified methods of cooperation on energy. In addition to direct purchases, Beijing negotiates with governments in the Gulf region to expand bilateral cooperation in exploration, boost refinery capacity, and attract the region's petrodollars to be invested in the refinery and petrochemical industry in China. The opening of the Fujian refinery/petrochemical manufactory in July 2005, jointly invested in by Saudi Aramco, Exxon-Mobil, and Sinopec, is a step toward the kind of economic cooperation and integration with the Gulf countries Beijing is hoping for.[40]

Finally, Beijing uses good relations with countries considered "problematic" by the West to open doors of opportunity for Chinese enterprises to do business. China's rhetoric of "non-interference" helps justify such practices. In Sudan, China's close governmental relations ensured that they were treated more favorably in their bid for contracts, while embargoes restricted Western companies from doing business. The first oil agreement with Sudan was signed in 1995, and was sponsored by the two governments.[41] As of June 2003, the CNPC had more investment in Sudan than in any other country in the world.[42] Because of its extensive oil interests in Sudan, Beijing has consistently opposed the UN taking strong measures against the Sudanese government.[43]

Beijing's consistent opposition to referring Iran to the UN over Iran's nuclear program is justified by similar rhetoric. In an interview given to a Chinese reporter on June 18, 2005, the Chinese ambassador to Iran criticized U.S. policy on Iran's nuclear issue as being based on "double standards."[44] China needs to be politically friendly with Iran to secure its $7 billion worth of annual trade and large oil contracts. After all, "the fruits of economic and trade cooperation between the two countries are being helped by favorable political relations," the Chinese commercial attaché to Iran said in 2005.[45]

Beijing's noninterference rhetoric is effective, and some in the region have explicitly attributed their willingness to do business with China to Beijing's non-interventionist approach to regional politics. An analyst at the Gulf Research Center in Dubai reportedly expressed the view that China's disinterest in forcing democracy in the region helped create a favorable business environment for itself.[46]

Offend-No-One Approach to Regional Politics

The second prong of China's strategy is to adopt an offend-no-one approach to regional politics. China's attitude toward regional conflicts in the past two decades has been consistently aimed at offending no regional or outside player. This has been achieved through rhetoric and ambiguous policies that do not commit Beijing to any party in a dispute. This is evidenced in China's policies toward the Arab-Israeli conflict, the Iran-Iraq war, the first Gulf war, and the recent war in Iraq.

To maintain good relations with both Israel and the Arab countries, Chinese rhetoric toward the Palestinian-Israeli conflict is more balanced today than in the past. During the Cold War, Beijing's effort to cultivate goodwill focused on strong rhetorical and material support for Arab nationalist causes against Israel.[47] This policy no longer serves China's economic interest in the region,[48] as the Chinese economy also benefits from Israel's advanced technology in areas such as agriculture and weapons manufacturing. Beijing established official diplomatic relations with Israel in 1992, and since then, China's attitude toward the Arab-Israeli conflict shifted from siding firmly with the Arabs to assuming the role of an impartial mediator.[49] On the one hand, Beijing insists on the principles of a two-state solution, "land for peace," and does not hesitate to condemn Israeli military actions in occupied territories. On the other hand, Beijing has reduced its rhetoric on the Palestinians' behalf, condemns terrorism, and urges the Palestinians to give up violence.[50] Wang Shijie, Beijing's first Middle East peace envoy, has visited the Middle East several times and met with the Israelis, the Palestinians, the Syrians, and the Egyptians. The envoy has not proposed any new ideas or any intention for Beijing to become more deeply involved in the peace process. Sending a Chinese peace envoy to the Middle East is a gesture that indicates a desire for peace in the region, as well as Beijing's balanced stance.

In an effort not to offend any one, Beijing's policies are often ambiguous and lack any principle other than "non-interference in others' internal affairs." During the long war between Iraq and Iran, China sold weapons to both countries without siding with either country. After Saddam's invasion of Kuwait, China refused to vote for a UN-sanctioned war on Iraq but condemned the invasion. Beijing opposed the recent war in Iraq but its criticism of the United States has been notably mild and Beijing has offered to provide aid and debt relief to the new Iraqi government.[51]

An overarching Chinese priority in the Middle East is to not be confrontational with the United States over regional issues. Beijing has made this decision because it realizes that its interests, both globally and in the Middle East and the Arab world, are best served by good relations with the United States.

In the region, China's economic interests require stability, which China does not have the resources to bring about by itself alone. It therefore relies on the United States as a primary force for peace and stability in the region.[52] Moreover, because of the United States' influence in the region, conflict with the United States would not only undermine Chinese diplomatic efforts and economic interests in the region, but also hinder China's global political and economic goals. First, such a conflict would inevitably damage Beijing's opportunity to work with the United States on priority issues like Taiwan. In other words, Beijing must pick its fights, and they are not in the Middle East and the Arab world. Second, conflict with the United States in the Middle East and the Arab world could be a potential impediment to China's pursuit of energy security in particular, and economic development in general. For example, Beijing is working to integrate itself into the international information- and technology-sharing networks to benefit from international cooperation on oil reserves and environmental protection.[53] China needs U.S. cooperation in order to join these networks, the most prominent of which is the International Energy Agency—the so-called "oil club" of the Western world.

This reasoning explains much of Chinese behavior in the region in the past decade. First, Beijing tried to highlight common interests and take measures in accordance with Washington's fight against terrorism and religious extremism. This has offered China an opportunity to increase coordination and information sharing with Washington throughout the Arab world. The Chinese emphasize that the two countries share common interests in maintaining stable oil prices and steady economic development. Second, Beijing has taken measures to alleviate tension with the United States over the issue of weapons proliferation. A number of times over the past decade, China has canceled sales deals with Iran, or made nonproliferation promises, out of a desire to please Israel or the United States.[54] China also swallowed the cancellation of Israeli sales of weapons under U.S. pressure.[55] Third, despite cozy relations with prewar Iraq and Iran's current leadership, as well as consistent opposition to either going to war in Iraq or referring Iran to the UN, Beijing has refrained from using its veto in the UN to oppose the policies of the United States.

Chinese analysts believe that the offend-no-one strategy has helped promote business in the region. For example, Zhu Weilie, director of the Middle Eastern Studies Institute of Shanghai Foreign Studies University, wrote that China's gaining of construction and oil contracts in Kuwait in 1992 was thanks to the Chinese government's "just and wise stance adopted during the Gulf crisis and the Gulf war."[56]

Soft Power

Culturally, exchanges and the rhetoric about the coexistence of civilizations are aimed at promoting goodwill favorable for business in the long term. The last few years have seen visibly increasing Chinese efforts to promote cultural understandings and inter-civilizational harmony. Cultural activities are aimed at generating goodwill and friendship. Rhetoric about inter-civilizational harmony seeks to distinguish China from the United States, and both help promote business.

To increase understanding and friendship, China is sponsoring more cultural activities in the Arab world, including the launch of several Chinese Cultural Weeks, Chinese Movie Weeks, and exhibitions of Chinese arts and history in Syria, Kuwait, Egypt, and elsewhere.[57] China is also funding Chinese cultural institutions in the Arab world. In October 2002, the first Chinese Cultural Center opened in Cairo, where locals can attend classes in Chinese language, martial arts, calligraphy, and dancing.[58] A Chinese-funded language school also opened in Egypt in 2003, and another in Damascus in March 2004. In December 2001, a nongovernmental Sino-Arab Friendship Association was established, headed by a Chinese Muslim and the deputy director of the standing committee of the National People's Congress, to encourage friendly activities and cultural understanding.[59]

In place of the outdated "third-world solidarity," the word "civilization" is being introduced into the vocabulary of friendship to cultivate goodwill in the region. The Chinese government claims to believe in the peaceful coexistence of civilizations and is unwilling to adopt the "clash of civilization" interpretation of world conflicts. First, Chinese leaders often tell visitors from the Arab world that the Chinese and Arab civilizations empathize with each other because both represent civilizations that have made great contributions to the world, and are facing the challenges of modernization and the encroachment of Western culture.

Despite improved relations with Israel and the West, Beijing still emphasizes its traditional friendship with the Arabs. For example, on the occasion of the establishment of the Sino-Arab Cooperation Forum in January 2004, Chinese Foreign Minister Li Zhaoxing said, "However the international situations change, the Chinese people will always be a sincere friend of the Arab people."[60]

Fully aware of Arab suspicion toward the secular nature of the Chinese regime, and despite its crackdown on Islamic militancy, Beijing has been trying to use the fact that China is home to 20 million Muslims to promote an image of itself as a friend of Islam. Since the 1980s, China has granted limited freedom for Islamic religious activities that have included educational exchanges and hajj visits by Chinese Muslims.[61] Arab visitors to China are routinely invited to meet with heads of the Islamic Association of China, visit Muslim regions of the country, and visit local mosques and Muslim families. Famous Chinese Muslim scholars, such as Na Zhong, who studied at the University of Azhar in Cairo in the 1930s, have been promoted as the "poster children" for Sino-Arab relations in the past.

Moreover, Beijing has appealed to Muslims by rhetorically delinking Islam from terrorism. When Chinese leaders express concerns to the Arabs over the threat of terrorism, they often emphasize that China is opposed to associating terrorism with certain people or religions (i.e., Arabs and Islam).[62] Similarly, Chinese analysts are critical of the view expressed by some Westerners that portrays Islam as the enemy.[63]

This three-pronged strategy reflects on China's foreign policy priorities and effective use of limited resources. China has substantial economic interests in the Middle East and the Arab world. The pursuit of these interests requires not only active diplomacy, but also efforts to increase Chinese influence in the region. Considering the uncertain future of the region, as well as China's limited hard power resources, Beijing has decided to quietly seek to expand its influence through exploiting soft power resources without antagonizing other players in the region, especially the United States.

Challenges to Beijing's Strategy

Beijing's strategy in the Middle East and the Arab world is paying off, as evidenced in the rapidly growing trade volumes, as well as the friendly relations with all players in the region. Beijing's strategy, however, faces challenges. First, some of China's policies and practices in the region have caused tension in U.S.-Chinese relations, and potential triggers for crises remain. Second, regional conflicts may force Beijing to choose sides between the United States and local actors, or between local actors themselves and the choices may produce unexpected or negative consequences. Third, issues such as human rights and separatism in Xinjiang undermine Beijing's efforts to cultivate goodwill and friendship.

Relations with the United States

Despite conscious efforts not to confront the United States, disagreements are inevitable—the most significant of which are issues of proliferation and Beijing's relations with "rogue" states. First, there are concerns that Chinese firms are directly involved in the proliferation of both conventional weapons and weapons of mass destruction, whether with or without the knowledge of the government. The U.S.-China Economic and Security Review Commission's 2005 annual report to Congress charged China with being "a source of WMD- and missile-related technologies to countries of concern such as Iran."[64] A 2006 Congressional Research Service (CRS) report expressed concerns over direct or indirect (through Pakistan) proliferation to Iran, Libya, and prewar Iraq.[65] Although the same CRS report also acknowledged the progress made by the Chinese government on the proliferation issue,[66] it stated that many believe "China can do and should be doing more to prevent the spread of WMD, missiles and conventional weapons."[67]

Second, policymakers in Washington are concerned about China's willingness to obstruct U.S. efforts to counter proliferation or deal with human rights issues in order to cultivate relations with energy-rich countries such as Iran and Sudan. The U.S. Department of Defense's 2005 report to Congress entitled "The Military Power of the People's Republic of China," charged that the pursuit of resources has brought China closer to problem countries such as Iran and Sudan.[68] Similarly, the January 2006 CRS report on China's proliferation activities revealed connections between China's opposition to referring Iran to the UN and China's energy deals with Iran.[69] Robert Zoellick, the U.S. deputy secretary of state, warned that "China might find its future energy security in peril by supporting Iran."[70] These issues, namely weapons proliferation and relations with rogue states, may generate potential triggers for crises in relations with the United States.

Relations with the Region

There are many possibilities for China to offend countries in the region. In the event of elevated conflicts, Beijing may be forced to choose sides. Moreover, Beijing's lack of concern for human rights may backfire and undermine China's public relations campaign. Finally, Beijing's treatment of its Muslim minorities may hurt its relations with the region.

Beijing's unwillingness to confront the United States prevents it from being fully supportive of its friends in the region. This failure to live up to the expectations of regional countries may disappoint China's friends and result in economic costs for Beijing. In the case of Iran, for example, the high priority given to good relations with the United States sets a limit on how long China can side with Tehran over the nuclear issue. At some point, Beijing may have to make a choice between alienating the United States and losing energy and construction contracts in Iran. However, because Beijing views good relations with Washington as critical to its development strategy it is highly unlikely that China will, for example, use its veto power in the UN Security Council on Iran's behalf.

Furthermore, at some point China may be forced to choose between countries in the region. For example, China's continuing support for Iran on nuclear issues may cause tension between China and Israel or the Arab states. Further, intensified conflict between Palestine and Israel may also force China off its current comfortable role as an impartial mediator.

Beijing's Effort to Cultivate Goodwill

Another problem in China's strategy is China's practice of ignoring human rights abuses. Although Beijing's tactic of justifying good relations with rogue states such as Sudan and Iran is generating short-term benefits, Beijing runs the risk of being viewed in the region as irresponsible or unconcerned with the implications of its policies. Political decisions driven by short-term interests may hurt China's image, and, as a result, also hurt its long-term economic interests in the region.

Separatism in China's northwest province of Xinjiang can be a potential cause for friction between China and the region. China is wary of Arab support for the Xinjiang separatist movement and related terrorist activities. Over the past two decades, the Chinese government relaxed control over foreign trade, foreign investment, and foreign travel regulations to attract Arab investment and promote trade between Muslim regions in the Northwest and the Muslim countries in Central Asia and the Arab world. As a result, Xinjiang and other predominantly Muslim provinces benefited from selling products, such as foodstuffs and clothes, to Arab countries and Iran.

With easier access to Xinjiang, Arab investment and funding of Islamic schools and institutions has increased. Arab buyers are visiting Chinese cities and towns, and some of them stay for long periods and even settle in China. Private connections between Arab and Chinese Muslims have grown, leading to consequences undesired by the Chinese government. For example, some Uyghur Muslims active in the separatist movement have received training in radical Islamic schools in countries such as Saudi Arabia and Syria. Throughout Chinese history, Muslim revolts against the central government have gained support from the Islamic world.[71] In the event of elevated conflict in Xinjiang, tensions could rise between China and Arab or Iranian governments. To date, however, Beijing has been quite successful at focusing relations on areas of mutual agreement.

Beijing's public relations campaign can be further obstructed because of its treatment of Muslim minorities in northwest China. The Qatar-based newspaper *Al-Jazeera*, for example, has reported a number of times on Muslim separatists in Xinjiang since its Hong Kong office opened. If the region democratizes further, freer media will openly report on problematic practices regarding Islam in China and hurt the image of China in the region. China's treatment of Muslims in Xinjiang threatens to undermine the goodwill China has accumulated over the years with the Arab world and Iran.

Conclusion

China's foreign policy in the Middle East and the Arab world focuses on using limited resources to secure energy supplies and expanding markets for Chinese products and services. To this end, China has adopted a three-pronged strategy. First, the state functions as an economic actor and uses diplomacy to advance Chinese economic interests in the region. Second, Beijing adopts an offend-no-one policy in regional politics to maintain an environment friendly to Chinese business interests. Third, Beijing exploits its soft power resources to cultivate goodwill in the region that will serve China's long-term economic interests.

China's strategy to pursue energy resources and markets in the region is not without challenges. China's goal to be friends with everyone may not always be achievable. First, some of China's policies and practices in the region have caused tension in U.S.-Chinese relations, and potential triggers for crises remain. Second, regional conflicts may force Beijing to choose sides between the United States and local actors, or between the local actors themselves, which may have unforeseen negative consequences. Third, Beijing's effort to cultivate goodwill and friendship can be undermined by issues such as human rights and separatism in Xinjiang.

In the long run, priorities can change. But in the short term, Beijing seems to have its priorities set. Beijing's strategy in the Middle East illustrates its general priority of using foreign policy to advance domestic economic goals. Out of concern for its own global and regional interests, Beijing is not currently willing to let its pursuit of economic resources in the Middle East and the Arab world damage its relations with Washington.

Notes

1. This chapter deals with the twenty-two members of the Arab League, as well as Israel and Iran. However, in the case of some North African states there is some overlap with the Joshua Eisenman's chapter on China-Africa relations. The members of the Arab League include: Egypt, Iraq, Jordan, Lebanon, Saudi Arabia, Syria, Yemen, Libya, Sudan, Morocco, Tunisia, Kuwait, Algeria, United Arab Emirates, Bahrain, Qatar, Oman, Mauritania, Somalia, Palestine, Djibouti, and Comoros.

2. Ministry of Commerce, Department of Western Asia and African Affairs, "Zhongguo-Alabo guojia hezuo luntan jianjie" (Brief introduction on the Sino-Arab Cooperation Forum), xyf.mofcom.gov.cn, March 27, 2006.

3. For works on the history of PRC-Middle East relations, see Yitzhak Shichor, *The Middle East in China's Foreign Policy, 1949–1977* (London: Cambridge University Press, 1979); Hashim S. H. Behbehani, *China's Foreign Policy in the Arab World, 1955–1975: Three Case Studies* (Boston: Kegan Paul International, 1981); John Calabrese, *China's Changing Relations with the Middle East* (London: Pinter Publishers, 1991); and Lillian Craig Harris, *China Considers the Middle East* (London: I.B. Tauris, 1993).

4. Pei Jianzhang, ed., *Zhong hua ren min gong he guo wai jiao shi, 1949–1956* (The diplomatic history of the People's Republic of China, 1949–1956) (Beijing: World Knowledge Publisher, 1993), 276.

5. Egypt, Syria, Yemen, Iraq, Morocco, and Sudan, despite opposition from the United States, became the first Arab countries to sever diplomatic relations with Taiwan, grant China recognition, and establish diplomatic relations in the second half of the 1950s.

6. Harris, *China Considers the Middle East*, 118.

7. For example, in 1965, China condemned Tunisian President Bourguiba for advocating peaceful coexistence with Israel. See Harris, *China Considers the Middle East*, 118.

8. See Joseph E. Khalili, *Communist China's Interaction with the Arab Nationalists since the Bandung Conference* (New York: Exhibition Press, 1970); also see Shichor, *The Middle East in China's Foreign Policy, 1949–1977*, 114, 124; also see Jiang Chun and Guo Yingde, *ZhongA guanxi shi* (Beijing: Jingji Ribao Publishing House, 2000), 297.

9. Guang Pan, "China's Success in the Middle East," *The Middle East Quarterly* (December 1997).

10. According to Abidi, China was buying oil from Iran to offset its sales of oil to Asian countries in the 1970s. Thus, in addition to good diplomacy in Asia, China also showed the Arab oil-producing countries the prospects of commercial deals.

11. "Minutes of Talks between Chou En-lai and Muhammad al-'Adsani, Kuwaiti Minister of Trade, Peking" (December 5, 1972); Behbehani, *China's Foreign Policy in the Arab World, 1955–1975: Three Case Studies*, 320–328.

12. Energy Information Administration, "China Country Analysis Brief" (August 2005). Available from www.eia.doe.gov/emeu/cabs/china.html, accessed April 28, 2006.

13. Yang Guang, "Cong nengyuan lianxi kan Zhongguo yu zhongdong guojia de huli hezuo" (Mutually beneficial cooperation between China and the Middle Eastern countries based on energy connections), *Xiya Feizhou,* no. 5 (October 2004): 53–58.

14. Wu Lei, "China's Oil Safety: Challenges and Counter Measures—with Concurrent Comment on the Influence of the Iraq War," *Xiya Feizhou,* no. 4 (August 2003): 17–21.

15. Wu Qiang and Qian Xuemei, "The Cooperation between China and the Mideast on Energy Affairs," *Strategy and Management,* no. 2 (April 1999): 49–52.

16. Zhu Weilie, "Alabo chanyouguo duihua nengyuan hezuo de guiji kaocha" (An international investigation on energy cooperation between Arab oil-producing countries and China), China Institute of International Studies. Available at http://www.ciis.org.cn/item/2005-06-29/51044.html.

17. "Foreign Cooperation Fruitful for Sinopec," *China Oil & Gas,* no. 4 (December 2002): 13–15.

18. Chinese Embassy in Libya, "Zhong shiyou (CNPC) zai Libiya di erlun EPSA shiyou kantan qukuai zhaobiao zhong zhongbiao" (CNPC won the second-round bid for exploiting Libya's EPSA fields) (October 5, 2005). Available from http://ly.mofcom.gov.cn, accessed November 2005.

19. Ministry of Commerce, Department of Western Asian and African Affairs (September 26, 2005). Available from http://xyf.mofcom.gov.cn.

20. Wu Qiang and Qian Xuemei, "The Cooperation between China and the Mideast on Energy Affairs," *Strategy and Management,* no. 2, (April 1999): 49–52.

21. Ministry of Commerce, Department of Western Asian and African Affairs (September 6, 2005).

22. Based on numbers provided on the Web site of the Ministry of Commerce. Available at http://xyf.mofcom.gov.cn.

23. Ministry of Commerce, Department of Western Asia and African Affairs, "2004 nian ZhongA maoyi tongji" (Statistics for Sino-Arab trade) (August 22, 2005). Available at http://xyf.mofcom.gov.cn/aarticle/zcfb/200508/20050800292372.html.

24. Interview with Chinese economic counselor at the Chinese Embassy in Iran (June 28, 2005). Available at www.southcn.com.

25. Ministry of Commerce, Department of Western Asia and African Affairs, "Zhongguo he Alabo guojia jingmao guanxi jiankuang" (Brief on the trade relationship between China and the Arab countries) (August 22, 2005). Available at http:/xyf.mofcom.gov.cn.

26. Ibid.

27. Based on numbers provided on the Web site of the Ministry of Commerce. http://xyf.mofcom.gov.cn.

28. Ministry of Commerce, Department of Western Asian and African Affairs, "Trade Statistics with Countries of West Asia and North Africa in 2005, January–December" (January 26, 2006).

29. The recent decision by Taiwanese President Chen Shuibian to cease the functioning of the National Unification Council was criticized by the Arab League. See "BBC Monitoring Asia Pacific-Political," *Lexis-Nexis* (March 4, 2006).

30. For example, the Ministry of Commerce decided earlier this year to use a special fund, the "Fund for promoting non-commercial overseas investment and trade exhibition," to help finance the 2005 Chinese Fair in Saudi Arabia to promote Chinese exports to Saudi Arabia, with which China has a huge trade deficit.

31. Responding to the gold rush in the Middle East and the Arab world, various government-owned research institutions and think tanks are engaged in research that provides information and analysis for both the government and business. Take *Xiya Feizhou* (West Asia and Africa), for an example. This journal, run by the Academy of Social Sciences, had over two dozen articles in 2004 on energy and trade issues in the Middle East and the Arab world. Issues concerning the authors of these articles range from economic reforms in the region, to U.S. sanctions toward Sudan, to exploring Iran's tourist market, and to the implications of the Iraq war on China's energy security.

32. "The Framework Agreement of the Cooperation in the Oil Sector between the Government of the People's Republic of China and the Government of the Islamic Republic of Iran," signed in Tehran on April 20, 2002.

33. "China Wants Closer Ties with Saudi Arabia, Gulf States, Premier Tells Visiting Saudi King." Associated Press Worldstream [Lexis-Nexis] (January 24, 2006).

34. "China Strengthens Mideast Oil Ties During Saudi's Visit," *The Seattle Times* [Lexis-Nexis] (January 25, 2006).

35. See CCTV interview with Sun Bigan, a former ambassador, charged with the effort to reestablish a Chinese embassy in Iraq. "Text of Interview Given to CCTV by Head of Group Charged with Rebuilding Chinese Embassy in Iraq," *Chinese Foreign Ministry* (February 17, 2004). Available at http://www.fmprc.gov.cn/chn/ziliao/wzzt/zt2004/xgxw/t66614.htm).

36. For example, the Chinese ambassador to Algeria, in a meeting with the new minister of State Affairs, spoke on behalf of the Chinese company responsible for the departure wing of the national airport in Algeria. He explained that the causes of the delay were many and responsibility fell not solely on the Chinese company. See Chinese Embassy in Algeria, "Zhu Aerjiliya dashi Wang Wangsheng baihui A guofang buzhang jian waijiao buzhang" (Chinese ambassador in Algeria, Wang Wangsheng, visits Algerian Minister of State Affairs and Foreign Minister) (June 9, 2005). Available at http://dz.china-embassy.org.

37. In another case, the Chinese Ambassador to Iran spoke at the Iranian-Chinese Chamber of Commerce and refuted the media claim that Chinese products are "cheap and of low quality," saying that individual Iranian businessmen were responsible for selling low-quality Chinese products. Chinese Embassy in Iran, "Liu dashi zai Yizhong shanghui 2004 nian gongzuo zongjie huiyi shang de jianghua" (Ambassador Liu's speech at the 2004 review meeting of the Iranian-Chinese Chamber of Commerce) (December 20, 2004). Available at http://ir.mofcom.gov.cn.

38. "Chief Executive of CNTIC Visits Chinese Ambassador in Iran" (May 27, 2005). Available at www.cntic.com.cn.

39. "The Association of Chinese-invested Enterprises in Iran Elects Its Second Board Members" (September 28, 2005). Available at http://ir.mofcom.gov.cn.

40. For more information on this project, see www.sinopec.com.

41. Zhu Weilie, "Alabo chanyouguo duihua nengyuan hezuo de guiji kaocha."

42. Alexander's Gas & Oil Connections, Company News, Africa, "CNPC and Sudan to Invest in Khartoum Oil Refinery," vol. 8, no. 12 (June 13, 2003). Available at www.gasandoil.com/goc/company/cna32441.htm.

43. For more information on Sudan, see chapter 2 on China and Africa.

44. "Zhuanfang Zhongguo zhu Yilang dashi" (A special interview with the Chinese ambassador to Iran) (June 2005). Available at http://www.ynfn.gov.cn/News/GMnew/200506/News_2519.html.

45. "Zhongyi jingmao hezuo shichu youdian leng, shiyou hezuo cheng zhongzhong zhizhong" (Sino-Iranian trade cooperation started cold, oil cooperation has become crucial). Available at http://www.southcn.com/news/china/china05/ses/wjky/200507070705.htm (June 28, 2005).

46. "China Strengthens Mideast Oil Ties During Saudi's Visit," *The Seattle Times* [Lexis-Nexis] (January 25, 2006).

47. China gained that reputation by providing Egypt with monetary donations and offering to send volunteers during the Suez crisis, supporting the Algerian independence war, and being one of the first countries to recognize the Algerian Liberation Front as the legitimate government of Algeria.

48. To use Algeria as an example, China provided generous military and financial aid including $10 million in credits in 1959 to the Algerian National Liberation Front (Harris, *China Considers the Middle East*,115). According to Yitzhak Shichor, Egypt, Syria, Yemen, and Iraq were the destinations of most of China's aid offers to the Middle East in 1956–76. In Shichor's account, the total aid offer to these countries in the period reached $381.1 million. See Shichor, *The Middle East in China's Foreign Policy, 1949–1977,* 209.

49. Militarily, Israel has become an important source of China's overseas weapons purchases. Since the cancellation of the Phalcon sale in 2000, Israeli arms deals with China continue to raise tensions between Israel and the United States.

50. For example, during his meeting with Yasser Arafat on April 15, 2000, Chinese president Jiang Zemin said that "whatever changes happen in the Middle East, Chinese policy to support the just cause of the Palestinians will not change." However, he stressed that force could not solve problems and advised Arafat to "follow the tide of the time" and adhere to peace negotiations. "Zhongguo he Balesitan dui fazhan weilai youhao guanxi de sidian gongshi" (Four mutual understandings between China and Palestine on a future friendly relationship), Xinhua News Agency (April 2000). Available at info.xinhuanet.com.

51. For example, see "China Decides to Forgive Iraq Debts, Reopen Embassy," *People's Daily* (February 6, 2004). Available at English.people.com.cn.

52. See an article written by two Chinese diplomats and former Chinese ambassadors to the Middle East, "Dui zhong dong he ping jin cheng de hui gu yu si kao" (Reflection on the Middle East peace process), *Wai jiao xue yuan xue bao* (Journal of Foreign Affairs University in Beijing), no. 3 (September 2003).

53. Wu Lei, "China's Oil Safety: Challenges and Counter Measures—with Concurrent Comment on the Influence of the Iraq War," *Xiya Feizhou,* no. 4 (August 2003): 17–21.

54. Congressional Research Service (CRS), "China and Proliferation of Weapons of Mass Destruction and Missiles: Policy Issues" (January 31, 2006); also see NTI, "China's Nuclear Exports and Assistance to Iran" (2003). Available at www.nti.org/db/china/niranpos.htm.

55. Shichor, *The Middle East in China's Foreign Policy, 1949–1977.*

56. Zhu Weilie, "Alabo chanyouguo duihua nengyuan hezuo de guiji kaocha."

57. For example, in January 2002, the Syrian cultural minister attended the Chinese fine art exhibition in Damascus. In February 2002, a "Chinese Movie Week" was held in Kuwait. In September 2002, the Syrian National Museum held an exhibition of Chinese antiques. In October 2003, another Chinese Cultural Week featured performances and exhibitions in both Cairo and Alexander. In March 2004, the Chinese Cultural Center in Egypt held a Chinese Cultural Week in Cairo and received wide coverage by local TV stations and newspapers.

58. "2002 nian Aiji yu Zhongguo guanxi dashi ji" (Important events in Sino-Egyptian relations during 2002), Xinhua News Agency (January 2003). Available at info.xinhuanet.com .

59. The Chinese government's desire to promote understanding of Arab culture and the fear of the spread of Islamic teachings is a constant battle. Chinese institutions often have to turn down donations from Arab governments, individuals, and religious organizations that have attached conditions aimed at Islamic proselytizing .

60. "Li Zhaoxing: Zhonga hezuo luntan de chengli juyou lichengbei yiyi" (Li Zhaoxing: The establishment of the Sino-Arab forum is a landmark), Xinhua News Agency (January 30, 2004).

61. Egypt's Azhar University, probably the highest institute for Islamic education in the Muslim world, has educated a number of Chinese Muslims sent by their government.

62. Song Gengyi, "Meiguo meiti dui yisilan shijie de xingxiang jiangou—yi shidai zhoukan gean weili" (American media's construction of the Islamic world: The case of *Time* magazine), *Arab World*, vol. 2 (April 2004): 9–11, at 11.

63. For other readings on Islam and the United States, see An Weihua, "Lun Mei guo yu yi si lan shi jie de chong tu" (On the conflict between America and Islam), *Studies of International Politics*, vol. 2, no. 88 (May 2003): 35–41.

64. U.S.-China Economic and Security Review Commission, "2005 Report to Congress." Available at www.uscc.gov.

65. CRS, "China and Proliferation of Weapons of Mass Destruction and Missiles: Policy Issues" (January 31, 2006): 11.

66. The CRS report says "China has taken some steps to mollify U.S. concerns about its role in weapons proliferation." CRS, "China and Proliferation of Weapons of Mass Destruction and Missiles: Policy Issues" (January 31, 2006).

67. International Information Programs, State Department, "China Must Enforce Its Nonproliferation Policies, Rademaker Says" (March 11, 2005). Available at http://usinfo.state.gov/eap/archive/2005/mar/14-967898.html.

68. Department of Defense, "Annual Report to Congress: The Military Power of the People's Republic of China, 2005," 2. Available at http://www.dod.mil/news/Jul2005/d20050719china.pdf.

69. CRS, "China and Proliferation of Weapons of Mass Destruction and Missiles: Policy Issues" (January 31, 2006): 11.

70. "Saudi Visit Underscores China's Growing Dependence on Mideast Oil," *Knight Ridder/Tribune News Service* [Lexis-Nexis] (January 25, 2006).

71. For example, Yakoub Beg, who successfully retained control over his rebel kingdom in Xinjiang for twelve years, received arms and the title "Commander of the Faithful" from the sultan of the Ottoman Empire.

6

China's Relations with South Asia

Rollie Lal

China has a long history of relations with India, beginning with cultural and religious contact between the two by 100 CE. Buddhism traveled from India through the Silk Route in Central Asia to China, mixing with the existing Daoist and Confucian philosophies there. The massive Himalayan mountain range, however, formed a natural barrier, limiting extensive communication between the two civilizations, as well as conflicts between the neighbors. But as travel and communication became easier, disagreements surfaced.

During the twentieth century, contact became far more frequent, and the emergence of independent India and Pakistan added complexity to these interactions, as did the Cold War and the interests of the Great Powers. After a short period of warm relations during the early 1950s, frictions over borders and geopolitical alignments strained the relationship. China developed close relations with Pakistan after 1971, providing political support and military assistance to India's foe. The growing ties between China and Pakistan led India to view both as collaborators against India's interests. A war between China and India in 1962 left both countries deeply suspicious of each other's intentions, and disagreements on Tibet and borders provide tangible cause for wariness.

After decades of tension, India and China are now moving closer and resolving some of their differences—a paradigm shift that could change the strategic realities of Asia. At the same time, Pakistan and India are making efforts to ameliorate relations with each other. Former Indian Prime Minister Vajpayee's visit to Beijing in 2003 produced a promising set of agreements to help settle the long-standing border dispute, increase trade, and decrease mutual distrust. This diplomatic initiative was received warmly, as both countries are moving toward a more interest-oriented approach to bilateral relations. A critical factor in drawing the two countries together has been their growing economies. However, while the probability of military conflict is declining, bilateral economic competition and geopolitics could be a new source of possible friction in the relationship.

China's relations with Pakistan, while still warm, now face a multitude

SOUTH ASIA
CHINA
MONGOLIA
INDIA
PAKISTAN
AFGHANISTAN
BURMA
THAILAND
VIETNAM
CAMBODIA
LAOS
SRI LANKA
MALDIVES
NEPAL
BHUTAN
BANGLADESH
NORTH KOREA
SOUTH KOREA
JAPAN
PHILIPPINES
INDONESIA
MALAYSIA
SINGAPORE
PAPUA NEW GUINEA
AUSTRALIA
NEW ZEALAND
SOLOMON ISLANDS
FEDERATED STATES OF MICRONESIA
MARSHALL ISLANDS
KIRIBATI
TUVALU
FIJI
VANUATU
SAMOA
TONGA
PALAU
EAST TIMOR
NORTH PACIFIC OCEAN
SOUTH PACIFIC OCEAN
INDIAN OCEAN
Arabian Sea
Bay of Bengal
Tropic of Cancer (23°27')
Tropic of Capricorn (23°27')
Equator

of new challenges. September 11, 2001 and the war in Afghanistan have increased the salience of terrorism for the United States and China. Pakistan's neighbors are increasingly concerned about the role of Pakistani radical groups as a destabilizing force in Afghanistan and China's western province of Xinjiang. Pakistan's lack of economic competitiveness relative to India makes it less attractive in a new Asia where economic realities determine geopolitical relations.

This chapter provides an overview of China's foreign policies toward India, Pakistan, and Afghanistan. It analyzes critical aspects of China's relations with these countries, focusing upon trends in the security and economic arenas.

Security Relations with India

Relations following revolution in China and independence in India were warm. A prevalent phrase reflecting the Indian view of China in the 1950s was "Hindi-Chini bhai-bhai," meaning that Indians and Chinese are brothers. The prime minister of India at that time, Jawaharlal Nehru, viewed Asia as a family of nations working together against imperialism. Revealing a worldview similar to that of contemporary Chinese leaders, he wrote, "[W]e stand for the freedom of Asian countries and for the elimination of imperialistic control over them."[1] Nonalignment would ensure the ability of each nation to pursue its interests without imperialist or superpower intervention. Peace as a strategy was made explicit in the concept of *panchsheel*, or the Five Principles of Peaceful Co-existence, which both China and India adopted during the 1955 Bandung Conference.

By the late 1950s, however, relations began to sour. Although China had assured India in earlier correspondence that the Tibet issue would be resolved by peaceful negotiations, China employed military force in 1950 to subjugate Tibet.[2] These actions alarmed the Indian government and set the stage for future Indian support of the Dalai Lama and the cause of the Tibetan people. In 1954, India recognized Chinese authority over Tibet in a bid to improve relations with China. The relations between the two countries began to deteriorate, however, and many Indians felt deceived by China.[3] In 1959, the Dalai Lama escaped an attack by the Chinese government and fled to India with a following of Tibetans. The Indian government provided the Dalai Lama immediate political asylum as well as permanent residence. From his base in Dharamsala in India, he worked for Tibetan freedom from China and protection of Tibetan human rights—an agenda looked upon with some anxiety by Beijing.

A poorly demarcated border between China and India caused further frictions. The McMahon Line, drawn by the British in 1914, was rejected by China, leading to prolonged territorial disputes. Existing tensions over Tibet and the dispute over the shared border finally led to a war between the neighbors in 1962. China's victory in the war did not resolve the territorial argument, and instead produced a legacy of mutual suspicion.

China's first nuclear test in 1964 also had an impact on Indian strategic thought. New Delhi saw those nuclear tests, closely following India's military defeat by China, as a strong argument for pursuing nuclear capability. Under Indira Gandhi, India performed its first nuclear test in 1974. Although described as a "peaceful" nuclear explosion, the tests were also a strategic response to rising tensions in the region. Deep-rooted suspicions that had grown between China and India over Tibet and boundaries were compounded by the realities of Cold War alignments.

In 1962–63, China and Pakistan reached agreements on border delimitations that indicated growing security collaboration against Indian interests.[4] Pakistan also played a critical role as the intermediary for U.S. President Richard Nixon's rapprochement with China. Prior to Nixon's presidency, close relations between China and Pakistan had posed a problem for the United States-Pakistan relationship. In contrast, President Nixon viewed the ties as an opportunity. In 1969, Nixon arranged for the establishment of secret communications with China via Pakistan—an association that led to the 1971 breakthrough in relations between the United States and China.[5] As China began its rapprochement with the United States, India developed warm relations and strong defense ties with the Soviet Union.[6] In the following years, Islamabad strengthened defense relations with Beijing, and benefited diplomatically from the partnership.

Despite these problems, both Zhou Enlai and Indira Gandhi moved to begin normalizing relations. Diplomatic relations were finally reinstated in 1976, breaking the tense impasse, and bilateral border talks were begun in the 1980s. In 1986, Indian military exercises in the border areas of Sumdorong Chu led to more turmoil. After Rajiv Gandhi's visit to China in 1988, a China-India Joint Working Group was established to expand the scope of the confidence-building measures then in place. After 1996, more measures were introduced. Border military meetings were held twice a year, hotlines were established between the two countries' militaries, each side notified the other in advance of exercises and movements, and a higher level of defense interaction and transparency in general was adopted.[7] Nonetheless, for decades both sides retained thousands of troops along their disputed borders. The end of the 1980s marked the beginning of renewed efforts on both sides to improve the bilateral relationship. Chairman of China's National People's Congress, Li Peng, visited India in 1991, and Beijing hosted Indian Prime Minister Narasimha Rao in 1993. Rao's visit produced significant progress, with the signing of the "Agreement to Maintain Peace and Tranquility along the Line of Actual Control." In 1996, Jiang Zemin visited India and both countries signed an agreement on additional military confidence-building measures, consolidating the trend of bonhomie.[8]

The Nuclear Tests

For Beijing, India's nuclear tests in May 1998 created a new set of tensions. Following the first round of tests, the Chinese response was muted, with only the foreign ministry expressing serious concern.[9] However, after India's second round, China condemned the Indian tests and criticized Pakistan's ensuing nuclear tests. China stressed that both India and Pakistan had undermined the international nonproliferation regime, and both should sign the Comprehensive Test Ban Treaty (CTBT) and Non-Proliferation Treaty (NPT) unconditionally. In May 1998, Indian Defense Minister George Fernandes said "China is potential threat number one," leading Beijing to interpret India's nuclear tests as a hostile maneuver aimed at China and exacerbating an already tense situation. The Chinese response turned sharply critical:

> The Chinese government is deeply shocked by this and hereby expresses its strong condemnation . . . The Indian government, which itself has undermined the international effort in banning nuclear tests so as to obtain the hegemony in South Asia in defiance of the world opinion, has even maliciously accused China as posing a nuclear threat to India. This is utterly groundless. . . . [T]his gratuitous accusation by India against China is solely for purpose of finding excuses for the development of its nuclear weapons.[10]

Chinese leaders insisted that China was not a threat to India, and therefore Indian nuclear disarmament was a necessity. Various scholars in China argued that the Indian claim of a Chinese threat was an excuse to garner sympathy from the West for its nuclear tests.[11] Some Chinese scholars argue, however, that the mixed reaction from China directly following India's tests reflected a lack of consensus among Chinese political and military leaders.[12] China stated that its strategy toward South Asia is one of cooperation, and that its military modernization is defensive and not directed at any particular country. Furthermore, the leadership noted that economic development is China's main priority, thus conflict is inimical to this national objective. Over time, the fact that neither India nor Pakistan is willing to forego its nuclear capabilities has become clear to the international community. China appears to have largely accepted the situation. The close military relationship between China and Pakistan complicates China's dealings on this issue. China's assistance to

Pakistan's nuclear and missile programs have undermined its credibility on nuclear nonproliferation. The close military relationship has also fueled Indian suspicions of a Chinese strategy of encirclement, creating further incentive for maintaining and expanding India's nuclear capabilities.

In March 2006, New Delhi took a big step forward on the nuclear issue, reaching an agreement with Washington that would allow India to purchase U.S. nuclear technology to bolster its civilian power industry. In return, India agreed to inspections of fourteen of its twenty-two nuclear facilities. Like Beijing, New Delhi is looking to significantly increase its nuclear power production, and in March 2006 had fourteen reactors in commercial operation and nine under construction.[13] China, however, expressed serious reservations about the deal, accused the United States of being soft on India, and warned of "a domino effect of nuclear proliferation [which] will definitely lead to global nuclear proliferation and competition."[14]

Recent Evolution of the Relationship

China and India have slowly come to an understanding that their national interests can be compatible. Both countries have committed themselves to the use of confidence-building measures and cooperative dispute resolution. China claims that India still occupies 90,000 square kilometers of Chinese territory in Arunachal Pradesh. India also accuses China of occupying 38,000 square kilometers of territory. However, discussions undertaken in 2003 helped to ameliorate relations. India moved to formally accept the Tibet Autonomous Region as a part of China—a long-standing sore spot. China also indicated flexibility on the issue of Sikkim. In the past, Chinese maps denoted Sikkim as a separate country, however China officially recognized Indian sovereignty over Sikkim in 2005 by presenting India with a map honoring New Delhi's claim.[15] Until October 2003, Chinese comments by a Chinese Foreign Ministry spokesperson remained ambiguous: "The question of Sikkim is one left over from history. We must respect history and take reality into consideration to address the issue. We hope that with the constant improvement of bilateral ties, the question will be solved gradually."[16] Despite any lingering disagreements, officials and scholars in both countries state that remaining border disputes are not worth war. As the populations of these two countries comprise a total of a third of the world, this is no small achievement.

In India, widespread support exists for the Tibetan people, their right to autonomy, and their right to their culture. However, the Indian government's position on Tibet—a largely accommodating one—is influenced heavily by national security and economic concerns. India can benefit from a reduction in China's military presence in Tibet, and from increased bilateral economic interaction. Despite official statements that India has accepted China's sovereignty over the Tibet Autonomous Region, the Tibetan issue remains contentious. The Dalai Lama continues to reside in Dharamsala, India, along with thousands of supporters. He remains spiritually influential and widely popular in India, creating a disparity between official actions and public sentiment. Many Indians disagree with the Indian government's stance on Tibet and support an independent, or at least autonomous, Tibet. In addition, strong popular opposition exists to China's policy of resettling ethnic Han in Tibet. The massive Han migration is rapidly changing the ethnic composition of the region, and will soon make Tibetans a minority in their own land.[17] Indians have also watched the destruction of various Tibetan Buddhist religious sites with dismay.

The U.S. role in Asia is also undoubtedly a significant factor in China's strategic calculus with regard to India. Soon after India's nuclear tests, China banded together with the United States to condemn India's nuclear program, creating apprehension in India that the United States and China might combine to restrain India. Conversely, calls within the United States to work with India to contain China have caused anxiety in China, providing Chinese strategists with incentives to warm relations with India. A closer relationship with India benefits China by precluding the United States from incorporating India into a containment strategy, whereas tensions between China and India would provide an opportunity for Washington and New Delhi to work together against China. Tensions between China and the United States over Taiwan give China even more incentive to resolve its disputes with India. China's primary geopolitical objective in Asia remains the favorable resolution of the Taiwan issue, thus the Chinese leadership is reluctant to expend military resources on territorial disputes to its south.[18]

China and India's growing economic and trade ties have contributed to an improving security relationship. Both countries have come to the understanding that economic cooperation is the key to the future, and closer diplomatic and security relations must follow. A senior military official in China noted, "I am not worried about relations between China and India because I think these two countries will come together to deal with the very similar challenges they will face in the future."[19]

The shift in relations was manifested in the first bilateral military exercises between India and China.[20] The exercises, held in November 2003, involved a joint naval search and rescue exercise off the coast of Shanghai and were symbolically important if limited in military significance. Military cooperation will undoubtedly decrease the possibility of miscommunication and misunderstanding, making conflict less probable. Joint exercises also serve to build confidence between the two countries' militaries, leading both sides to claim they were building a "strategic partnership" in April 2005. However, China will take time to deepen its security ties with India, and the exercises do not signal the imminent start of a military alliance or an extensive amount of trust. Economic cooperation will serve to bring Beijing and New Delhi closer together, even as it adds some new competitive elements into the mix.

Evolving Economic Relations with India

Commerce has been the main driving force in the new Sino-Indian bilateral relationship. China's economic success since it began "reforming and opening" in 1978 set a positive example for India. New Delhi only responded a decade later, however, after an economic crisis forced India to reform its bureaucratic and largely socialized economic system. Liberalization and growth has decreased the incentives for both China and India to engage in conflict with each other, and increased the initiatives to reconcile differences.

Despite progress, India's GDP was approximately $720 billion in 2005 —less than half of China's ($1.79 trillion).[21] And, although on the rise, India's incoming foreign direct investment—up 41 percent in 2005–6 to $7.5 billion—also remains low compared with China's $60.3 billion in 2005 and $14.25 billion in the first quarter of 2006.[22] Because India implemented reforms a decade after China, it will take time for India to match China's success. India, however, does have advantages over China. The banking sector is considerably healthier than China's. Insolvency is approximately 15 percent, whereas China's insolvency is estimated to be between 25–40 percent.[23] India's growth has been largely internally driven, with exports comprising only 19 percent of GDP in 2004 (34 percent for China). This low export dependence makes the economy less vulnerable to fluctuations in international demand or other external shocks. The service sector also comprises a larger part of India's economy. The World Bank estimated that 53.2 percent of India's GDP was derived from services, while 40.7 percent of China's economy was service driven in 2004.[24] Services are the present engine of growth for the Indian economy in the absence of a robust manufacturing sector, and provide a stable engine for India's growth. As the manufacturing sector grows, India has the capability of further increasing economic growth.

There is likely to be tremendous growth in trade between the two countries in the coming years. Trade volume increased from a paltry $260 million per year in 1990 to $18.6 billion in 2005, and both countries agreed to boost bilateral trade to $20 billion by 2006.[25] This trade provides a boost to both economies and increases interdependence. Of course, economics also provides a new field for competition. India eyes with envy China's rapid growth rates and competitiveness in the consumer goods sector. Indian leaders had worried that cheap Chinese goods were flooding the Indian market and threatening domestic manufacturing. In the past few years, however, businesses are looking to China more as a model for attracting foreign investment. Indian leaders businesses realize that much effort is needed to match China's performance in this arena, including added attention to education, infrastructure, and less bureaucracy.

China, for its part, is hoping to emulate India's success in the information technology arena. In China, recent media reports indicate rising concern regarding India's increasing competitiveness in microchip manufacturing. Chinese analysts argue that because India's salaries are lower, costs are cheaper, thereby making Indian products more competitive. Language is also a factor. Chinese businessmen fear that U.S. businesses will prefer Indian products because of the Indian facility with English. In a strange turn of events, the Chinese population is now asking whether their market is likely to be flooded with cheap Indian goods. With common strengths and export markets, trade competition is inevitable. But competition on economic terms is certainly preferable to competition in the security arena, and beneficial to both.

Factors Drawing the Two Countries Together

Both countries have entered a higher level of development, and progress from here requires a strict focus on economic policies. Beijing and New Delhi are addressing their challenges with national policies intended to propel their economies and simultaneously strengthen their militaries. Both face the threat of growing income disparity, ethnic unrest, and separatism, and each is focused on modernizing and developing its economy.

Both countries also share several similar interests and goals vis-à-vis the international system, including: territorial sovereignty, military development, environmental standards, human rights standards, and access to energy. The interest in territorial sovereignty applies to Taiwan, Tibet, and Xinjiang for China, and to Kashmir for India. Paradoxically, agreement over the principle of territorial sovereignty brings both together, but differences over specific territorial questions still present points of contention. Military development and modernization remain key objectives for both countries in their pursuit of becoming stronger players on the world stage. China and India share a critical interest in ensuring that environmental and human rights standards for their own development remain at the levels that the developed countries faced. In addition, high tensions with Taiwan provide China with a good reason to resolve old quarrels with India.

A critical difference is that Indian policy makers are more confident than their Chinese counterparts regarding their ability to deal effectively with domestic ethnic and economic forces. Indian leaders overwhelmingly state in interviews that the unity of the Indian state does not hinge upon keeping Kashmir, whereas in the view of most Chinese policy makers, a separation from Taiwan could mean the end of China as we know it.[26] Rather than concern itself with national unity, India is trying to refocus its national efforts on economic growth to match China's success. Though Indian growth rates have averaged 6–8 percent in the past decade, growth needs to be even faster to eradicate poverty and raise living standards. Conflict and tensions with neighboring China and Pakistan have posed a large economic hurdle for India in the past, impeding foreign investment by increasing risk, and absorbing critical budgetary resources. Recent diplomatic overtures may alleviate this problem and allow China and India, in particular, to focus on their rapidly expanding economic relationship.

China's View of a Rising India

Chinese leaders and scholars have not considered India a peer competitor for decades, but that may now be changing. China's reforms in the 1980s, and the ensuing rapid growth in the 1990s, allowed it to surge ahead of India as an economic powerhouse in Asia. The military capabilities of China were rapidly advancing, as China acquired nuclear weapons as well as the ability to deliver them to all parts of India.

China's membership in the UN Security Council also accorded it a status that India could not attain (although in 2005, Beijing did formally support New Delhi's campaign for membership).[27] At the same time, Chinese leaders were cognizant of the fact that India could eventually become a competitor—an eventuality that made the leadership uncomfortable.

India's development of a long-range, nuclear-capable ballistic missile has the potential to bring all of China into range. The Indian Navy's acquisition of an aircraft carrier creates concerns among Chinese defense strategists, as does the increasing effectiveness of India's Air Force.[28] In addition, India's rapid economic growth since the mid-1990s and its competitiveness in the information and service sectors have drawn the attention of Chinese leaders. Amelioration of ties has been the primary policy, which could deter India from aligning with the United States to contain China while providing a mutually beneficial economic relationship. Nonetheless, as India continues to grow, if China shows reluctance to accord the country status as a peer in the international arena, the resulting friction could lead to renewed tensions.

China and Ties with Pakistan

While China's relationship with India has generally been growing closer over the last several years, its relationship with Pakistan has been growing more distant. During the Cold War and the 1990s, China maintained a close relationship with Pakistan—to India's dismay. During the 1960s, China emerged as the major weapons supplier to Pakistan, and the defense relationship was cemented over the following decades. Pakistan's dispute with India over Kashmir had its benefits for China. A critical military route from Xinjiang through Tibet in Kashmir was ceded by Pakistan to China, although the region remained under dispute with India.[29] The route provided logistical and military access for China into Tibet.

One area where Sino-Pakistani ties remain robust is military assistance. In April 2005, Beijing sold Islamabad four F-22P naval frigates and six Z-9C helicopters designed for "surface-to-surface and surface-to-air missiles along with numerous associated self-defense systems." The agreement also involves training of Pakistan naval personnel and the transfer of technology.[30]

China's continued military assistance to Pakistan, including the transfer of nuclear and missile technology, has been a persistent irritant in Sino-Indian relations. In the 1980s and 1990s, China supplied F-5A and F-7 fighters to Pakistan, as well as several hundred tanks, anti-tank missiles, and other conventional weapons.[31] China provided the short range, nuclear-capable DF-11 or M-11 missile to Pakistan. Declassified documents indicate that Chinese nuclear transfers to Pakistan spurred consternation in the U.S. government for decades. In 1983, U.S. State Department analysts concluded that China was helping Pakistan with fissile material production, and 1992 papers revealed concern regarding China's "continuing activities with Pakistan's nuclear weapons programs."[32] In 1995, Pakistan procured approximately 5,000 ring magnets for uranium enrichment centrifuges from China.[33] Despite this, China continued to deny reports of these sales, stating after the 1998 tests that it had not transferred any technology to Pakistan that could be used for the manufacture of nuclear weapons.[34] Revelations of Pakistani nuclear and missile sales to Libya, North Korea, and Iran indicate that Pakistan's nuclear programs, made possible only with Chinese help, have now taken on a life of their own. Chinese sales of nuclear technology to Pakistan might have ended years ago, but they have enabled Pakistan to provide Chinese nuclear technology to other countries. And despite official denials of nuclear proliferation, it is possible that China had advance knowledge of the international transfers. This is a particular concern given numerous past statements that China would not allow transfers of nuclear technology to a third country without China's prior consent.[35] In the 2004 investigations following the revelation of Pakistani nuclear transfers to Libya, Libyan scientists provided early Chinese nuclear weapons designs wrapped in bags and tagged with a Pakistani address to the IAEA.[36]

China has assisted Pakistan heavily in the civilian nuclear arena as well. The first nuclear plant built in Chashma, Pakistan by China became operational in 1999. Currently, China is engaged in building a second nuclear power plant for Pakistan in Chashma, known as C-2. The plant is expected to be complete in six years.[37] China is also assisting Pakistan in the development of the Gwadar Deep Sea Port in the province of Baluchistan, funding approximately 80 percent of the project. When completed next year, Gwadar is expected to be Pakistan's third largest port, providing Arabian Sea access to Central Asia, Afghanistan, and Xinjiang and shortening the distance from Xinjiang to the sea by approximately 2,500 km.[38] Hundreds of Chinese workers are engaged in building the port. In May 2004, a terrorist bombing in Gwadar killed three of the Chinese workers—the first attack on Chinese nationals in Pakistan.[39] The incident drew international attention to the project and speculation about China's intentions in the Indian Ocean. Indian analysts fear Chinese entry into the Indian Ocean, whereas the United States is apprehensive of the proximity to Diego Garcia. Most recently, both countries have increased the level of military interaction through joint exercises. In August 2004, China and Pakistan embarked upon the first joint antiterrorism exercises between the two countries. Border troops from each country met in the Pamir Mountains of Xinjiang to engage in exercises simulating attacks upon terrorists.

However, China has in recent years emphasized its intent to pursue a balanced foreign policy toward India and Pakistan, effectively downgrading its relationship with Pakistan. During the 1999 Kargil conflict between India and Pakistan, China notably did not support the Pakistani incursion, emphasizing the necessity to resolve the Kashmir dispute through negotiations rather than military means. This shift is likely a result of India's growing significance as an economic and military power in Asia.

Other issues are also increasingly affecting China's relations with Pakistan. Revelations of Pakistan's transfer of nuclear technology to North Korea placed China in a difficult position vis-à-vis the international community and North Korea. Initially, China found that it had to scramble to defuse the situation in Northeast Asia created by North Korea's move to nuclear capability. China ultimately refused to implicate Pakistan, however, questioning the U.S. sources of information regarding Pakistani nuclear proliferation and the origin of North Korea's nuclear program.[40]

In addition, whereas Pakistan's support for fundamentalist groups was previously not a priority for China, reports now indicate that the number of Uyghur separatists trained by Pakistan has created problems for China in Xinjiang. As a result, the issue has become a liability for Pakistan in the bilateral relationship.[41] Nonetheless, the strong historical relationship between the two countries is unlikely to disappear. Military cooperation continues, and China may want to maintain close ties with Pakistan as a hedge against being surrounded by a hostile United States, Japan, and India in an unknown future.

Security remains the focus of the China-Pakistan relationship and accounts for the majority of interactions. Despite efforts to foster broader ties, economic interests are a small component of bilateral relations—in 2005, the total bilateral trade volume was about $ 4.26 billion, compared to $ 18.6 billion with India.[42] Agricultural products form the bulk of Pakistan's exports to China, whereas manufactured products are generally imported by Pakistan.[43] Pakistan's weak manufacturing base is an impediment to growth in the export trade of higher value-added goods to China, as Chinese products compete favorably against Pakistani goods.

Despite financial constraints in Pakistan that pose a hurdle to expanded economic growth, both countries continue to emphasize the need to expand economic relations. In April 2005, Chinese Premier Wen Jiabao visited Islamabad to establish a free trade agreement between the two nations. The agreement included provisions to expand Chinese investment and boost Pakistani agricultural imports by applying special tariff arrangements to 767 products. Under the agreement, all exportable items from Pakistan, including textile goods, surgical and sporting goods, vegetables, fruits, rice, citrus, and mangoes, would have market access in China with tariffs reduced to zero from January 2006.[44]

China and Afghanistan

Afghanistan borders China's western province of Xinjiang—a region populated by a Muslim and ethnically Turkic group known as the Uyghurs. China has had historical problems in politically and culturally integrating the Uyghurs, and the proximity of Xinjiang to Muslim Afghanistan and the Central Asian republics has complicated the issue. Uyghurs groups have frequently resisted Chinese rule since the first Chinese takeover of the region in 1759, and remain a challenge for the Chinese government.

With the defeat of the Soviet Union in Afghanistan, and the subsequent independence of the Central Asian states, the Uyghurs gained inspiration for their own independence movement. A separatist organization, the East Turkestan Islamic Movement (ETIM) formed and many militants began training alongside other Islamic militants in Afghanistan.[45] According to the Chinese government, hundreds of Uyghurs were trained there for terrorist activities.[46] Uyghur separatists have been charged with launching attacks both in Xinjiang and across China's borders in Central and South Asia. Taliban support for China's Uyghurs also created a paradox in China's foreign policy. Despite the fact that China remained close to Pakistan, Pakistan had become the strongest supporter of the Taliban, the primary group that was assisting the Uyghurs. Over time, this situation likely caused relations between China and Pakistan to erode, despite Pakistan's official change of policy toward the Taliban after September 11, 2001.

The continuing assistance of the Taliban to the Uyghurs in recent years has spurred the Chinese government to take a more intense interest in Afghanistan. Since the overthrow of the Taliban regime, China has been active in helping to reconstruct Afghanistan, pledging $150 million in assistance over five years to the country in 2002. In 2004, China also agreed to cancel all the Afghan government's old debts (dating back to 1965), estimated to be approximately $18 million.[47] Projects for rebuilding the Jamhoriat Hospital in Kabul and an irrigation system in Parwan province are also underway. Chinese workers in Afghanistan, however, have been the targets of terror attacks, with eleven construction workers killed in June 2004 in an attack by twenty gunmen.[48] The increase in terror attacks against Chinese nationals in Pakistan and Afghanistan, combined with the insurgency in Xinjiang, has led to a growing interest in counterterrorism by the Chinese government. In the past, Islamic fundamentalism in areas other than Xinjiang was not a high priority for the Chinese government, although it was accepted as a plausible tool for destabilizing countries viewed as a threat. Now, as the linkages between radical groups in different countries become clear, the Chinese government may take an interest in stronger measures to manage the problem outside its borders.

Conclusion

China's relations with the countries of South Asia have become extremely complex in recent years. Whereas relations with Pakistan have been consistently warm for decades, interactions with India appear to be moving in a more positive direction. Growing economic ties with India are providing "Cold Warriors" on both sides of the border incentives to work out past differences. After a long period of tension, even joint military exercises have become possible.

While the apparent warming of ties between China and India does not necessitate a cooling between China and Pakistan, the trilateral dynamic has changed. China's economic ties with Pakistan have grown rapidly in recent years, but they remain a fraction of the booming trade between China and India. China has also undertaken a more balanced policy toward India and Pakistan, shifting away from its former policies of siding with Pakistan on the Kashmir dispute. In doing so, China has indicated that positive relations with India may increasingly be more critical to its interests than maintaining the status quo with a smaller and weaker Pakistan.

Recent diplomatic initiatives with India are likely to push forward this trend in Chinese foreign policy. At the same time, U.S. interests in Asia also create conditions for China to move closer toward India. Chinese leaders are apprehensive that a United States-Indian alignment could form to contain China. India's growth as an economic power in Asia and a military competitor make the scenario of containment even more formidable. In an effort to preempt any such possibility, Chinese leaders are fostering closer diplomatic relations with India.

Nonetheless, disagreements between China and India remain in critical areas such as Tibet and Chinese military assistance to Pakistan. These issues will not be resolved swiftly or with ease. China's close military ties with Pakistan are a component of a broader security relationship between the two countries that is meant to protect each country's geopolitical interests in the region. Furthermore, as China and India grow as peer competitors in Asia, both countries will increasingly compete for leverage in the international arena. India may demand equal footing with China, whereas China may attempt to contain India. Casting new roles in the international system for these massive countries will be a challenge for the international community, and will be critical to creating long-term stability in Asia.

Notes

1. Jawaharlal Nehru, *India's Foreign Policy: Selected Speeches, September 1946–April 1961* (Bombay: Publications Division, Ministry of Information and Broadcasting, Government of India, 1961), 24.

2. Chih H. Lu, *The Sino-Indian Border Dispute: A Legal Study* (Westport: Greenwood Press, 1986), 49–53.

3. John W. Garver, *Protracted Contest: Sino-Indian Rivalry in the Twentieth Century* (Seattle: University of Washington Press, 2001), 52.

4. Dennis Kux, *India and the United States: Estranged Democracies* (Washington, D.C.: National Defense University Press, 1992), 210.

5. Ibid., 279–281.

6. John W. Garver, "The China-India-U.S. Triangle: Strategic Relations in the Post-Cold War Era," *NBR Analysis*, vol. 13, no. 5 (October 2002).

7. These measures were adopted in the "Agreement on Establishment of Confidence-Building Measures in Military Field along the Line of Actual Control in the India-China Border Area." See Xia Liping, "China-India Security Relationship: Retrospects and Prospects," Shanghai Institute for International Studies, at www.siis.org.cn/english/collection/xialiping.htm, accessed February 22, 2005.

8. See Garver, *Protracted Contest*.

9. "China Seriously Concerned About Indian Nuclear Test," Xinhua News Agency (May 12, 1998).

10. "Chinese Government Strongly Condemns Indian Nuclear Tests," Xinhua News Agency (May 14, 1998).

11. Zou Yunhua, "Chinese Perspectives on the South Asian Nuclear Tests," *Working Paper* (Palo Alto, CA: Stanford University's Center for International Security and Cooperation [CISAC], January 1999).

12. Ming Zhang, *China's Changing Nuclear Posture: Reactions to the South Asian Nuclear Tests* (Washington, D.C.: Carnegie Endowment for International Peace, 1999), 30.

13. "US and India Seal Nuclear Accord," *BBC* (March 2, 2006). Available at http://news.bbc.co.uk/2/hi/south_asia/4764826.stm.

14. Mohan Malik, "China Responds to the U.S.-India Nuclear Deal," *Jamestown China Brief,* vol. 6, no. 7 (March 29, 2006). Available at www.jamestown.org/publications_details.php?volume_id=415&issue_id=3670&article_id=2370926.

15. John Lancaster, "India, China Hoping to 'Reshape the World Order' Together," *Washington Post* (April 12, 2005). Available at www.washingtonpost.com/wp-dyn/articles/A43053-2005Apr11.html.

16. "Foreign Ministry's Spokesperson's Press Conference on 14 October," *ChinaView* (October 18, 2003). Available at www.xinhuanet.com/english/briefings.htm, accessed August 26, 2004.

17. "Tibet: Chinese Pouring into Lhasa," *New York Times* (August 8, 2002); "China Revives Controversial Tibetan Migration Project," Agence France-Presse (January 23, 2002).

18. Chinese military officials and scholars, in interviews with the author, 2001–2003.

19. Chinese military official, in a conversation with the author, Beijing, China.

20. "Naval Exercises Indicate Sea Change in Relations," *South China Morning Post* (November 15, 2003); "First India-China Naval Exercise Opens New Chapter in Ties," *The Press Trust of India* (November 15, 2003).

21. *CIA World Factbook.* Available at www.cia.gov/cia/publications/factbook/index.html, viewed on May 7, 2006.

22. "Foreign Direct Inflows at Record $ 7.5 Billion," *DNAIndia* (April 20 2006). Available at www.dnaindia.com/report.asp?NewsID=1025112.

23. James Kynge, "China's 'Big Four' Banks Set for New Loan Plan," *Financial Times* (February 5, 2004); "China Economic Review: China Mulls Multi-Bln-Dlr Bailout Plan on State Bank," Xinhua News Agency (January 17, 2005); Sonia Kolesnikov-Jessop, "China Soft Landing Has Cost," United Press International (July 19, 2004); Khozem Merchant, "India Set to Launch Bank Recovery Vehicle: A Plan to Restructure Distressed Public Sector Debt Could Face Legal and Political Hurdles," *The Financial Times* (April 10, 2002).

24. World Bank, *World Development Indicators* database (2006). Available at www.worldbank.org/data/countrydata/countrydata.html.

25. "China-India Bilateral Relationship," Embassy of the People's Republic of China in India. Available at www.chinaembassy.org.in/eng/focus/t247108.htm; "India-China Bilateral Trade Touches USD 13.6 billion," *The Press Trust of India* (February 6, 2005); "Sino-Indian Trade Tops US $10 billion," *People's Daily Online* (December 28, 2004). Available at http://english1.people.com.cn/200412/28/eng20041228_168942.html.

26. Data from interviews with Indian and Chinese policy makers and scholars, 2001–2003.

27. Lancaster, "India, China Hoping to 'Reshape the World Order' Together."
28. Interviews with Chinese policy makers and scholars, 2001–2003.
29. Garver, *Protracted Contest*, 205.
30. "China to Build Four Frigates for Pakistan Navy" (April 5, 2005); "Pakistan, China Finalizes Frigates' Deal" (posting on the PRC Embassy in Pakistan's Web site, July 7, 2005). Available at http://pk.china-embassy.org/eng/zbgx/t203371.htm.
31. Garver, *Protracted Contest,* 235.
32. Siobhan McDonough, "Documents Show U.S. Unease Over Pakistan-China Security Cooperation," Associated Press (March 5, 2004).
33. Daniel L. Byman and Roger Cliff, *China's Arms Sales: Motivations and Implications* (Santa Monica: RAND, MR-1119-AF, 1999).
34. Garver, *Protracted Contest,* 222–223.
35. "China's Nuclear Exports and Assistance," *China Profiles* database, Center for Nonproliferation Studies, Monterey Institute of International Studies. Available at www.nti.org/db/china/nexport.htm.
36. "China's Nuclear Exports and Assistance to Pakistan," *China Profiles* database. Available at www.nti.org/db/china/npakpos.htm.
37. "China Signs Deal to Build Second Nuclear Plant in Pakistan," *AFX-Asia* (May 5, 2004).
38. "China-Built Pakistan Port 'Draws Animosity' from India, U.S.," *Straits Times* Web site, *BBC Worldwide Monitoring* (May 17, 2004).
39. "Bodies of Chinese Car Bomb Victims Flown to China from Pakistan," Agence France-Presse (May 7, 2004).
40. Joseph Kahn and Susan Chira, "China Challenges U.S. on Pyongyang's Arms," *International Herald Tribune* (June 10, 2004).
41. "Hong Kong Daily Cites Report on Uighur Muslim Leader Put to Death in China," *South China Morning Post* Web site, *BBC Monitoring International Reports* (October 24, 2004); Richard McGregor, "Uighur Training Angered Beijing," *The Financial Times* (October 18, 2001).
42. "Hu in Pakistan to Expand Bilateral Trade, Friendship," *China Daily*. Available at www.chinadaily.com.cn/china/2006-11/24/content_741607.htm.
43. "Pakistan, China Pledge to Set up Free Trade Area," *Business Daily Update* (December 29, 2004).
44. "Pak, China Sign Treaty of Friendship; Beijing's Assurance to Defend Territorial Integrity, Sovereignty," *The Dawn* [Karachi, Pakistan] (April 6, 2005). Available at http://skyscrapercity.com/archive/index.php/t-199968.html.
45. Paul George, "Country Perspectives China: Islamic Unrest in the Xinjiang Uighur Autonomous Region," *Commentary No. 73, Canadian Security Service Publication* (Spring 1998).
46. "China Demands Return of Nationals Caught Fighting for Taliban," Agence France-Presse (December 11, 2001).
47. "China Offers Aid, Debt-Relief to Afghanistan," *China Daily* (April 2, 2004); "China to Write Off Huge Afghan Debt," Xinhua News Agency (March 27, 2004).
48. "China Says 10 Chinese Construction Workers Killed in Afghanistan," Associated Press (June 9, 2004).

7

Stabilizing the Backyard

Recent Developments in China's Policy Toward Southeast Asia

Michael A. Glosny

In testimony before the Senate Foreign Relations Committee in 2005, U.S. Assistant Secretary of State for East Asian and Pacific Affairs Christopher Hill said, "China's most dramatic diplomatic, political, and economic gains of the past few years have been in Southeast Asia."[1] Leaders and analysts in the United States, China, and the countries in the Association of Southeast Asian Nations (ASEAN) have also observed that one of the most successful elements of Chinese diplomacy in recent years has been its successful strategy toward Southeast Asia, which has produced remarkable improvements in Sino-ASEAN cooperation.[2] This chapter analyzes the development of China's policy toward Southeast Asia, particularly since the Asian financial crisis of 1997, and discusses why Sino-ASEAN relations have improved so dramatically in a relatively short period of time.[3]

In an attempt to reassure countries along its periphery, Chinese leaders have made references in recent years to "becoming friends and partners with neighbors" (*yulinweishan, yilinweiban*), building an "amicable, tranquil, and prosperous neighborhood" (*mulin, anlin, fulin*), and have even characterized China as a "friendly elephant" (*youhao de daxiang*) in a Sino-ASEAN meeting.[4] Chinese statements expressing trust and friendship toward its neighbors are not new. Even China's attempts to interfere in Southeast Asian countries during the 1960s and 1970s by fostering the spread of revolution, which led to many years of mistrust and suspicion from Southeast Asian countries, were often accompanied by friendly rhetoric.[5] What is new, however, is that since the mid- to late 1990s, China has consistently taken action to back up these words of friendship and goodwill in its relations with ASEAN countries.[6] In the early to mid-1990s, China responded to Southeast Asian leaders who voiced concern about the possible negative implications of China's economic growth and growing power by condemning such claims as groundless, or suggesting that they were the result of the influence of the Western media's fictitious and

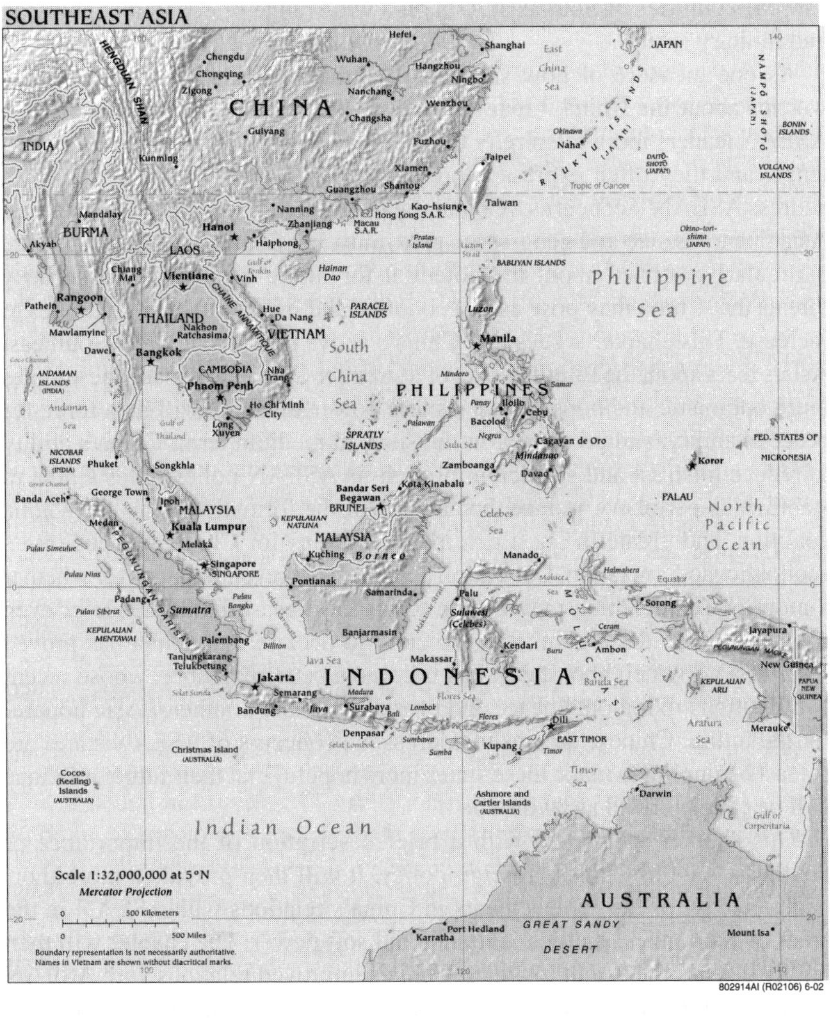
SOUTHEAST ASIA
CHINA
Hefei
Shanghai
Chengdu
Wuhan
Hangzhou
Ningbo
Chongqing
Zigong
Nanchang
Wenzhou
Changsha
Guiyang
Fuzhou
Kunming
Taipei
Xiamen
Guangzhou
Shantou
Taiwan
Kao-hsiung
Nanning
Hong Kong S.A.R.
Zhanjiang
Macau S.A.R.
Hanoi
Haiphong
Pratas Island
Luzon Strait
HENGDUAN SHAN
INDIA
Mandalay
BURMA
Akyab
LAOS
Gulf of Tonkin
Hainan Dao
Chiang Mai
Vientiane
Vinh
Rangoon
Pathein
Mawlamyine
THAILAND
Nakhon Ratchasima
Hue
Da Nang
PARACEL ISLANDS
VIETNAM
CHAINE ANNAMITIQUE
Dawei
Bangkok
CAMBODIA
Phnom Penh
Nha Trang
Ho Chi Minh City
Long Xuyen
South China Sea
ANDAMAN ISLANDS (INDIA)
Andaman Sea
Gulf of Thailand
NICOBAR ISLANDS (INDIA)
Phuket
Songkhla
Banda Aceh
George Town
Ipoh
MALAYSIA
Medan
Kuala Lumpur
Melaka
Pulau Simeulue
Pulau Nias
Singapore
SINGAPORE
Padang
Sumatra
PEGUNUNGAN BARISAN
Pulau Siberut
KEPULAUAN MENTAWAI
Palembang
Pulau Bangka
Tanjungkarang-Telukbetung
Jakarta
Bandung
Semarang
Java
Surabaya
Madura
Bali
Lombok
Denpasar
Sumbawa
Flores
Sumba
Kupang
Timor
Dili
EAST TIMOR
Christmas Island (AUSTRALIA)
Cocos (Keeling) Islands (AUSTRALIA)
Indian Ocean
KEPULAUAN NATUNA
Kuching
Borneo
Pontianak
MALAYSIA
BRUNEI
Bandar Seri Begawan
Kota Kinabalu
Samarinda
Banjarmasin
Java Sea
Makassar
INDONESIA
Palu
Sulawesi (Celebes)
Kendari
Manado
Celebes Sea
Sulu Sea
Zamboanga
Cagayan de Oro
Mindanao
Davao
SPRATLY ISLANDS
Palawan
Bacolod
Negros
Panay
Iloilo
Cebu
Samar
Mindoro
PHILIPPINES
Manila
Luzon
BABUYAN ISLANDS
Philippine Sea
East China Sea
JAPAN
RYUKYU ISLANDS (JAPAN)
Okinawa
Naha
NAMPŌ SHOTŌ (JAPAN)
BONIN ISLANDS
DAITŌ-SHOTŌ (JAPAN)
VOLCANO ISLANDS
Tropic of Cancer
Okino-tori-shima (JAPAN)
FED. STATES OF MICRONESIA
Koror
PALAU
North Pacific Ocean
Halmahera
Molucca Sea
MOLUCCAS
Equator
Sorong
Ceram
Buru
Ambon
Banda Sea
Jayapura
PEGUNUNGAN MAOKE
New Guinea
PAPUA NEW GUINEA
KEPULAUAN ARU
Merauke
Arafura Sea
Timor Sea
Ashmore and Cartier Islands (AUSTRALIA)
Darwin
Gulf of Carpentaria
AUSTRALIA
GREAT SANDY DESERT
Port Hedland
Karratha
Mount Isa
Scale 1:32,000,000 at 5°N
Mercator Projection
500 Kilometers
500 Miles
Boundary representation is not necessarily authoritative.
Names in Vietnam are shown without diacritical marks.
802914AI (R02106) 6-02

exaggerated "China threat theory." By the mid- to late 1990s, Chinese leaders and analysts began to fear that China's assertive foreign policy was beginning to encourage some ASEAN countries to join a coalition to balance Chinese power and potentially hinder China's development.[7] Shortly thereafter, China began proposing initiatives and adopting policies that were designed to reassure the countries of Southeast Asia on a wide range of economic, political, and military issues.

As one measure of how successful China's ASEAN policy has been, worries about the China threat have rarely been openly voiced since 1997. ASEAN leaders more often refer to China as both an opportunity and a challenge, and they often express appreciation for China's concrete actions to address ASEAN's concerns.[8] China already has a large impact on Southeast Asia due to its size and geographic proximity, but Southeast Asian states are particularly worried about the potential for future economic and security threats that China may pose as its economic and military power continues to increase. This chapter argues that China's active policy to address Southeast Asia's fears about the future impact of a stronger China have made these states more optimistic and hopeful that a more powerful China will be a force for peace, stability, and prosperity in the short to medium term. China's ability to reduce mistrust and suspicion through its ASEAN policy helps China to stabilize its periphery, reduce the chances of the formation of an anti-China coalition, and create the best external conditions for China to focus on its economic modernization. Long-term concerns, however, remain over whether a more powerful China may decide to try to increase its influence or even try to dominate the region. It is too early to declare that China has proved itself as a good neighbor, but it has become a better neighbor, whose recent contributions to regional peace and prosperity make Southeast Asia hopeful for the future. Chinese actions to address the concerns of ASEAN states are at least beginning to make these states more hopeful that their future relations will be characterized as "win-win."

This chapter will begin with a brief description of the importance of Southeast Asia in Chinese foreign policy. It will then provide a more thorough analysis of the achievements in China's relations with ASEAN in the areas of economics, politics, military, and soft power. The chapter will then briefly discuss the extent to which China's improved relations with ASEAN are allowing China to achieve a dominant position in Southeast Asia. The pessimists on this issue overestimate China's influence, underestimate the continued strong position of other regional actors such as the United States and Japan, overlook the extent to which China's recent activities have also indirectly served U.S. interests, and ignore the obstacles that China will need to overcome in its future relations with ASEAN. Lastly, I will briefly examine some of the key obstacles that China needs to overcome if its relations with ASEAN are going to continue to improve.

Chinese Foreign Policy and Southeast Asia

Southeast Asia is an area of vital importance to China due to traditional security concerns and reform-era economic interests. China's history of being invaded by powers along its periphery has produced a deep-seeded, almost obsessive, fear of invasion and encirclement.[9] Although China's traditional concerns were focused on foreign forces to the north and west, these concerns expanded with the arrival of European navies in the nineteenth century to include the possibility of invasion from Southeast Asia.[10] From China's perspective, the United States and the Soviet Union both used Southeast Asia as an important component in their plots to encircle China during the Cold War. These perceived security threats on China's border made the leadership of the People's Republic of China (PRC) feel that it needed to respond militarily, and China's leaders were willing to make economic sacrifices to ensure that the PRC had sufficient military capabilities to deal with such threats. After the collapse of the Soviet Union and its withdrawal from Southeast Asia, China tried to maintain good relations with the region to remove any incentive for Southeast Asia to join in an anti-China coalition. Given America's strong relationship with Japan, several Chinese analysts refer to Southeast Asia as the best place to break the U.S. strategic encirclement (*dapo Meiguo zhanlue weidu*) of China.[11]

Stabilizing China's periphery has and will continue to allow China to focus on economic development during what China's leadership has proclaimed the "period of strategic opportunity" (*zhanlue jiyuqi*) in the early twenty-first century.[12] As a source of natural resources, investment, and trade, Southeast Asia is, and will continue to be, an important driver of China's economic modernization. Regional stability is also important because the majority of China's trade, including oil imports, passes through maritime Southeast Asia. China also hopes that increased economic interaction with continental Southeast Asia can help fuel economic growth in China's economically backward southwestern provinces, such as Yunnan and Guangxi. China's improved relations with Southeast Asia may also help further isolate Taiwan and make it less capable of achieving permanent separation, if not formal independence. Regional trade and security agreements that do not involve Taiwan limit the island's ability to be a part of regional cooperation mechanisms, and as Southeast Asian countries value their relations with the PRC more and more, they will become less likely to offend the mainland by supporting Taiwan's political maneuvers.

Many Chinese analysts also contend that improving relations with its periphery will help China hedge against a possible downturn in U.S.-China relations, which many analysts fear will follow the current honeymoon period.[13] The most effective way to deal with foreign pressure, according to this view, is to build a "ring of political friendship" on China's periphery.[14] If China improves its relations with countries on its periphery, this may also allow China to command more respect and be in a stronger position to handle its relationship with the United States.[15]

Finally, some have tied the need to improve relations with regional countries to China's desire for Great Power status, arguing that all Great Powers have had a strong footing in their own regions. According to this view, if China hopes to exercise power and influence in Asia or the world, it must be able to do so in its own backyard.[16] It is too early to tell whether a greater focus on peripheral countries reflects China's long-term intentions to outmaneuver Japan and establish its own sphere of influence in East Asia that excludes the United States, or if it reflects a desire to increase intraregional economic cooperation and political stability solely for the sake of China's economic modernization.

Addressing Economic Concerns: Helping Southeast Asia Achieve Growth in the Twenty-First Century

The Asian financial crisis and pressure from China's continued economic growth have recently combined to push ASEAN to be more concerned with domestic political stability and economic growth and less concerned with external security threats. By the late 1990s, Southeast Asia's fears of a "China threat" usually referred to China's economic growth coming at the expense of other countries in the region. In an effort to reduce these fears, China has used rhetoric and specific policy proposals to help address the specific concerns of ASEAN countries about financial stability, trade, and investment. Chinese leaders have tried to emphasize the benevolent nature of China's sacrifices in many of these economic proposals. By emphasizing "win-win" (*shuang-ying*) outcomes and adopting policies that address regional economic fears, China's actions have helped make ASEAN leaders more optimistic that a stronger China may prove to be a driver of economic growth and prosperity in Southeast Asia, while also reducing overall political mistrust.

Coping with Globalization: ASEAN Plus Three and a New Regional Architecture

The Asian financial crisis, which hit Thailand in the summer of 1997 and spread throughout Southeast Asia, showed that initial moves toward establishing an ASEAN Free Trade Area had done little to shield these countries from the dangerous fluctuations in exchange rates and interest rates that came with globalization. China's short-term assistance during the financial crisis showed that it could play a constructive role in helping Southeast Asia weather the storm. Its active involvement in long-term financial monitoring and assistance programs also showed China's devotion to preventing the recurrence of such a crisis in the future.

China's contribution to assistance efforts during the financial crisis, both through the International Monetary Fund and through bilateral loans and aid, was greater than most ASEAN countries had expected.[17] ASEAN was even more appreciative of China's steadfast refusal to devalue the Renminbi, which would have set off another round of competitive devaluations and done more damage to ASEAN. Chinese officials tried to characterize it as an example of a responsible and unselfish power putting itself at a disadvantage for the sake of others, but avoiding a deepening of the financial crisis was also in China's interests as it would have hurt China's economy as well.[18] According to then Secretary-General of ASEAN, Rodolfo Severino, "China is really emerging from this smelling good."[19]

After the United States insisted on attaching onerous conditions to aid and opposing the establishment of an Asian Monetary Fund during the financial crisis, East Asian countries established a new discussion mechanism—the ASEAN Plus Three (APT)—to develop regional solutions to future problems that did not rely on assistance from the United States.[20] The first APT summit took place in December 1997 in Kuala Lumpur, and included discussion on finance and monetary issues. These annual meetings led to the creation of the APT Surveillance Process and the APT Early Warning System, which are aimed at improving regional monitoring capabilities and information sharing to reduce the risks of a future financial crisis.

In May 2000, in what came to be known as the Chiang Mai Initiative, the APT finance ministers agreed to the establishment of bilateral swap agreements to increase the region's ability to cope with potential future liquidity shortfalls.[21] Since 2000, APT members have entered into sixteen different bilateral swap arrangements totaling $35.5 billion, and in May 2005 they agreed to double the total amount of these swap arrangements.[22] Even after the acceleration of these swap arrangements in 2005, only 20 percent of the total available amount can be withdrawn without an International Monetary Fund program being in place, which severely limits the potential for autonomous assistance. There were also early moves toward the establishment of an Asian bond market in 2003 and 2004.

China opposed the establishment of an Asian Monetary Fund in 1997, because the proposed fund was a Japanese initiative and China felt relatively insulated from speculative attacks. Beijing later warmed to such initiatives.[23] Although the results are still relatively limited in scope, China's active participation in changing the regional financial architecture has helped deepen Sino-ASEAN relations and shown that China is willing not only to take short-term action to help ASEAN economies, but also to work to redesign the regional economic architecture to better protect ASEAN from external economic threats.

Competing with China: Addressing ASEAN's Concerns over Trade Competition

ASEAN states are worried that with China's low-cost and increasingly efficient manufacturing sector, cheaper Chinese goods will flood their domestic markets and compete effectively with Southeast Asian-produced goods in other markets.[24] In recent years, for instance, ASEAN has been losing market share in the United States and Japanese textile and clothing markets, and worries that as safeguard quotas continue to be lifted, ASEAN will lose out on a more level playing field.[25] Rather than attempting to convince ASEAN countries that these worries are unfounded, Chinese officials are openly acknowledging them, and have proposed economic policies that would help address ASEAN's specific worries, improve ASEAN's economic situation, and demonstrate that China's growth is not necessarily threatening to ASEAN.[26]

China's economic growth has led to tremendous increases in Sino-ASEAN bilateral trade, and an ever-expanding market for ASEAN's products. As Table 7.1 illustrates, increases in bilateral trade have been astonishing, including a fourfold increase in total trade between 1997 and 2004. During President Hu's April 2005 visit to Indonesia, he called for passing $200 billion in total Sino-ASEAN bilateral trade by 2010.[27] Although bilateral trade started at a very low level—$7.28 billion in 1990, and China still conducts only a small percentage of its overall trade with ASEAN—these increases in bilateral trade have helped deepen economic integration, and have demonstrated to ASEAN that China's economic success can help power the ASEAN countries' own economies. China has also endured a large trade deficit with the ASEAN-6 countries that has, in part, helped these states recover from the financial crisis by exporting to the Chinese market (see Table 7.2). This has led to further explanations of China's benevolence toward ASEAN and its sincere desire for a win-win situation.

In the context of discussions over how ASEAN would cope with China's accession to the WTO, Chinese Premier Zhu Rongji surprised ASEAN by proposing the establishment of an ASEAN-China Free Trade Area (ACFTA) at the November 2000 APT summit. ASEAN and China signed an agreement

Table 7.1

Recent Growth in Sino-ASEAN Total Trade

	1997	1998	1999	2000	2001	2002	2003	2004
Total trade in billions of US dollars (% change)	25.1	23.6 (–5.9)	27.2 (15.3)	39.4 (44.9)	41.8 (6.1)	54.8 (31.1)	78.2 (42.7)	105.9 (35.4)
% of PRC's total trade conducted with ASEAN-10	7.7	7.3	7.5	8.3	8.2	8.8	9.2	9.2

Source: International Monetary Fund, *Direction of Trade Statistics Yearbook*, 2005.

Table 7.2

China's Balance of Trade with ASEAN-6 (net bilateral trade in billions of U.S. dollars)

1990	1.02
1994	300
1997	–.930
1998	–2.95
1999	–3.63
2000	–5.50
2001	–6.04
2002	–9.51
2003	–19.2
2004	–23.1

Source: IMF, *Direction of Trade Statistics*, various years.

to implement the ACFTA at the November 2002 APT summit. If the ACFTA is smoothly implemented, tariffs and non-tariff barriers should be removed for ASEAN-6 (Brunei, Indonesia, Malaysia, Philippines, Singapore, and Thailand) countries by 2010 and for CLMV (Cambodia, Laos, Myanmar or Burma, and Vietnam) countries by 2015. ACFTA will create a free trade area of 1.7 billion people, with a combined GDP of $2 trillion and total trade exceeding $1.2 trillion. According to simulations, it should increase ASEAN's exports to China by 48 percent and increase China's exports to ASEAN by 55 percent, leading to an increase in ASEAN's GDP of 0.9 percent and an increase in China's GDP by 0.3 percent.[28] At the 2004 APT summit in Vientiane, the parties signed more agreements showing that the ACFTA is on its way toward successful implementation.[29]

Table 7.3

China's Balance of Trade with CLMV (net bilateral trade in billions of U.S. dollars)

1990	.2
1994	.4
1997	1.2
1998	1.4
1999	1.0
2000	.7
2001	1.4
2002	1.9
2003	2.8
2004	3.0

Source: IMF, *Direction of Trade Statistics,* various years.

By allowing ASEAN to benefit from preferential import duties before other WTO members, the ACFTA will help alleviate ASEAN's worries over trade competition by giving it special access to the Chinese market.[30] China also agreed to "early harvest" arrangements, which would reduce the duties on imports of 600 agricultural goods for those countries like Thailand that wished to liberalize even quicker. One Chinese scholar suggests that the early harvest arrangement (*zaoqi shouhuo*) is "an example of China's flexible (*linghuo*) position in the negotiations."[31] Another refers to the "concessions" China made in the terms offered to ASEAN.[32] Most Chinese analysts and leaders focus on the win-win and mutually beneficial nature of the arrangement, and emphasize the costs China is willing to bear to address ASEAN's worries and help drive further economic growth.

Although China has been running trade surpluses with CLMV countries (see Table 7.3), it has used other tools to help fuel economic growth in these countries. China has become more involved in the Greater Mekong Subregion Economic Cooperation Program, established by the Asian Development Bank in 1992 to promote economic integration between the CLMV countries, Thailand, and China. China has also invested in new roads and rail lines, taken part in the Cross-Border Transport Agreement that streamlines customs and immigration procedures, and hosted summits such as the annual China-ASEAN Expo to attract investment. From 1999 to 2002, China invested an annual average of about $80 million dollars in CLMV countries, which usually accounted for a majority of China's foreign direct investment to ASEAN countries.[33] In 2005, China donated $20 million to the Asian Development Bank (ADB) to set up a regional poverty reduction center that will place special emphasis on assisting Indochina.[34] At the 2005 APT summit, Premier Wen Jiabao also promised to

Table 7.4

Investment Diversion in Sino-ASEAN Relations

	% of FDI inflow into developing Asia-Pacific absorbed by ASEAN-5	% of FDI inflow into developing Asia-Pacific absorbed by China
1990	31.2	15.6
1996	27.5	42.8
1997	25.2	41.8
1998	19.4	44.4
1999	23.0	35.7
2000	14.5	27.8
2001	15.5	41.9
2002	12.6	55.8
2003	13.3	52.7
2004	15.7	40.6

Source: For 1990 data, see UNCTAD, *World Investment Report, 1996*; for 1996–1997 data, see UNCTAD, *World Investment Report, 2002*; for 1998–2002 data, see UNCTAD, *World Investment Report, 2004*; for 2003–4 data, see UNCTAD, *World Investment Report, 2005.*

provide more than $3 billion in economic assistance and concessional credit over the following three years, with most going to CLMV countries.[35] Lastly, China gave the CLMV countries until 2015 to comply with the terms of the ACFTA, allowing them to take advantage of China's domestic market. Fueled by China's assistance, the subregion's GDP is expected to rise from $212 billion in 1997 to $863 billion in 2010.[36] Chinese leaders hope that this active involvement will help stabilize its southern periphery and increase the level of economic development of its poorer southwest provinces.

Losing out to China: Addressing ASEAN's Concerns over Investment Diversion

ASEAN countries are also concerned that as China develops, it will attract much of the investment that used to go to the ASEAN countries—severely damaging ASEAN's future manufacturing capabilities. Although investment figures are particularly difficult to measure, most studies show a clear trend of investment diversion from ASEAN to China since the 1997 financial crisis (see Table 7.4).[37] Chinese scholars and leaders have focused their energies on convincing ASEAN that Chinese economic growth will lead to more future investment in ASEAN, showing how China's growth will drive ASEAN's growth.

As China's economy continues to grow, Zhang Yunling explains that "ASEAN will become the first choice of where to invest for Chinese companies."[38] Another scholar argues that as Sino-ASEAN trade continues to increase, Chinese companies will want to locate plants closer to their markets in Southeast Asia.[39] At the 2003 APT summit in Bali, Premier Wen Jiabao declared China would make an active effort to increase foreign direct investment (FDI) in ASEAN, and said that such investment was growing at an annual rate of 20 percent, and in some countries by as much as 40 percent.[40]

Although announcing plans to focus on increasing FDI in ASEAN may receive political support from ASEAN, China's FDI is so small that it will be unlikely to help the more advanced ASEAN economies for some time. Between 2000 and 2002, China's annual FDI to ASEAN was only approximately $100 million to $200 million.[41] Although China's investment in some sectors such as energy has skyrocketed—reaching $1.2 billion in Indonesia—it is not clear why China would invest in ASEAN's manufacturing and industrial sectors if low production costs continue to make China's domestic environment more favorable.[42] After ACFTA's full implementation, ASEAN's unfettered access to the China market might attract investment, but it may take many years to develop.[43] For the short term, China has focused on the promise of future investment in ASEAN as a solution to ASEAN's investment shortfalls, and another example of how China's growth will help drive growth in Southeast Asia.

Addressing Political Concerns: Slowly Rebuilding Trust between Neighbors

Due in part to the historical legacy of China's Cold War policy of interfering in Southeast Asia's internal affairs and trying to spread revolution, the most fundamental problem in post-Cold War Sino-ASEAN relations has been, and will continue to be, a lack of mutual trust. Although China's ability to project power beyond its borders is still limited, ASEAN states are suspicious of how a more powerful China may behave toward its neighbors. ASEAN states, especially those in Indochina, worry about future Chinese influence, control, and domination. By becoming more supportive of regional multilateral organizations, and increasing the interactions with ASEAN leaders through mutual visits, Chinese leaders hope they will be able to reduce suspicion over China's future intentions and convince ASEAN leaders that a rising China will be responsible and non-threatening.

China's Changing Evaluation of Regional Multilateral Organizations

In the uncertain environment immediately following the Cold War, every ASEAN state wanted to engage China, especially through multilateral organizations. They calculated that this would at least provide information about Chinese capabilities and intentions while constraining aggressive behavior, and might also help to socialize China into becoming a more responsible, peaceful, and cooperative power. Although China joined the Asia-Pacific Economic Cooperation (APEC) forum in 1991, China was more anxious about joining an organization devoted to political and security issues. Chinese leaders feared that such institutions would allow the Great Powers to set an agenda designed to contain China's growth, allow smaller powers to unite and gang up against China, and offer a forum for interference in its internal affairs. They also feared that these institutions would try to internationalize the Taiwan issue and other territorial disputes.[44] China joined the ASEAN Regional Forum (ARF) when it was formally established in 1994, but only because it was reassured that ASEAN's leading role in the ARF—including the adoption of the "ASEAN Way" of informality and consensus decision-making—meant that the ARF would not move too fast or push China too far.[45]

Although China's decision to join the ARF showed its desire to be a part of regional organizations, Chinese leaders dragged their feet in the first few years of their involvement in the ARF and prevented any real progress.[46] Today, however, China actively engages in virtually every regional multilateral organization, including the ARF, APEC, APT, the Council for Security Cooperation in the Asia-Pacific (CSCAP), Asian Cooperation Dialogue, ASEM (Asia-Europe Meeting), and the Forum for East Asia-Latin America Cooperation (FEALC). By the early twenty-first century, China was not only actively involved with these organizations, it became eager to host summits, and even started proposing the establishment of new multilateral organizations. China and ASEAN also signed the Joint Declaration on Strategic Partnership for Peace and Prosperity in October 2003, and China acceded to the Treaty of Amity and Cooperation in October 2003. China's shift in attitude toward regional multilateral organizations, and its public respect for the region's values, have made it easier for ASEAN leaders to believe that China may play a more constructive role in the region and be somewhat constrained in taking actions harmful to the region's interests.

Joint Declarations, Cooperation Mechanisms, and Mutual Visits

After laying the foundation for a stronger relationship with ASEAN as an organization, Chinese leaders turned their attention to doing the same in its bilateral relations with each ASEAN member nation. Between February 1999 and November 2000, China signed joint statements with each ASEAN member that laid out basic principles both sides would follow to improve relations in the future, and made explicit reference to China's Five Principles of Peaceful Coexistence and ASEAN's Treaty of Amity and Cooperation. Each declaration emphasized working to increase trade and investment, building political cooperation through mutual high-level visits, strengthening defense cooperation, and included a reaffirmation of the one-China policy.[47] China followed up these joint declarations by arranging repeated high-level mutual visits to ASEAN countries (see Table 7.5).

When China and ASEAN signed the Joint Declaration on Strategic Partnership for Peace and Prosperity in October 2003, it showed just how far these relations had come. According to the 2003 joint declaration, "the relationship between ASEAN and China has seen rapid, comprehensive and indepth growth and ASEAN and China have become important partners of cooperation" since 1997.[48] Rather than speaking in vague generalities about the hopes of a better relationship in the future, as the 1997 joint declaration had done, this document was very specific in describing the various mechanisms for cooperation that had developed, and the accomplishments and agreements that were in the process of being implemented. Although it is difficult to evaluate the impact of these visits and agreements, they showed that China was willing to focus much attention, at multilateral and bilateral meetings, to improving its relations with ASEAN.

A Political Milestone: Signing ASEAN's Treaty of Amity and Cooperation

In terms of reducing ASEAN's political mistrust toward China and building trust and goodwill, signing ASEAN's Treaty of Amity and Cooperation (TAC) in October 2003 was probably China's greatest achievement. ASEAN Secretary General Ong Keng Yong called it "trailblazing," and several Chinese leaders referred to it as a "milestone" (*lichengbei*).[49] Signatories of the Treaty of Amity and Cooperation, which dates from 1976, should be guided by principles such as the mutual respect for the independence, sovereignty, equality, and territorial integrity of nations, as well as noninterference in the internal affairs of others, peaceful settlement of disputes, and renunciation of the use of force. Agreeing to follow these principles, and taking actions consistent with them, was especially meaningful for ASEAN states and made China's promises to uphold the principles of sovereignty and noninterference in the internal affairs of others much more believable.[50] ASEAN leaders have also given China much credit for starting this chain reaction that led to India, Japan, South Korea, and Russia all subsequently signing it. This shared vision of how states should interact with each other has helped reduce mistrust in Sino-ASEAN relations.

Table 7.5

Important Recent High-Level Visits in Sino-ASEAN Relations

Date	Official	Destination
Feb. 2002	President Jiang Zemin	Vietnam
Mar. 2002	Brunei Crown Prince Muhtadee Billah	China
Mar. 2002	Indonesia President Megawati Soekarnoputri	China
April 2002	Vice President Hu Jintao	Malaysia
April 2002	Vice President Hu Jintao	Singapore
May 2002	Foreign Minister Tang Jiaxuan	Indonesia
Sep. 2002	Chairman of NPC Li Peng	Indonesia
Sep. 2002	Chairman of NPC Li Peng	Philippines
Feb. 2003	Thai Prime Minister Thaksin Shinawatra	China
April 2003	Gen. Sec. of Vietnamese Communist Party Nong Duc Manh	China
June 2003	Foreign Minister Li Zhaoxing	Vietnam
July 2003	Foreign Minister Li Zhaoxing	Singapore
Sep. 2003	Malaysian Deputy PM Abdullah Badawi	China
Oct. 2003	President Hu Jintao	Thailand
May 2004	Singapore Deputy Prime Minister Lee Hsien Loong	China
May 2004	Vietnam Prime Minister Phan Van Khai	China
May 2004	Malaysian Prime Minister Abdullah Ahmad Badawi	China
June 2004	Vietnam Foreign Minister Nguyen Dy Nien	China
June 2004	Foreign Minister Li Zhaoxing	Philippines
July 2004	Myanmar Prime Minister Khin Nyunt	China
Sep. 2004	Philippines President Gloria Macapagal-Arroyo	China
Oct. 2004	Premier Wen Jiabao	Vietnam
Oct. 2004	Cambodian King Norodom Sihamoni	China
Nov. 2004	Former Foreign Minister and special envoy Tang Jiaxuan	Indonesia
Nov. 2004	Former Foreign Minister and special envoy Tang Jiaxuan	Thailand
Nov. 2004	Former Foreign Minister and special envoy Tang Jiaxuan	Malaysia
Jan. 2005	Singapore Foreign Affairs Minister George Yeo	China
Feb. 2005	Malaysian Supreme Head of State Tuanku Syed Sirajuddin	China
Mar. 2005	Philippine Secretary of Foreign Affairs Alberto G. Romulo	China
April 2005	President Hu Jintao	Brunei
April 2005	President Hu Jintao	Indonesia
April 2005	President Hu Jintao	Philippines
May 2005	Chairman of NPC Wu Bangguo	Singapore
May 2005	Chairman of NPC Wu Bangguo	Malaysia
July 2005	Thai Prime Minister Thaksin Shinawatra	China
July 2005	Vietnam President Tran Duc Luong	China
July 2005	China Foreign Minister Li Zhaoxing	Laos
July 2005	China Foreign Minister Li Zhaoxing	Myanmar
July 2005	Indonesian President Susilo Bambang Yudhoyono	China
Aug. 2005	Cambodian King Norodom Sihamoni	China
Oct. 2005	Singapore Prime Minister Lee Hsien Loong	China
Nov. 2005	President Hu Jintao	Vietnam
Dec. 2005	Premier Wen Jiabao	Malaysia
Feb. 2006	Myanmar PM Soe Win	China
April 2006	Premier Wen Jiabao	Cambodia
April 2006	Singapore Senior Minister Goh Chok Tong	China

Source: Center for Strategic and International Studies, *Comparative Connections*, various issues.

Addressing Military Concerns

In the early to mid-1990s, China's rising military budgets, in combination with little defense transparency, outstanding territorial disputes, and a demonstrated willingness to use force to settle these disputes, constituted a major potential threat to Southeast Asian nations. China's lack of significant power projection capability, the focus on Taiwan, deft diplomacy, and the region's recognition that China is placing priority on attending to its severe domestic challenges (particularly economic development) have somewhat reduced fears of Chinese aggression in Southeast Asia over the short term. However, China's military modernization remains a potential long-term worry, especially as the ASEAN states recognize that military equipment purchased to be used against Taiwan could also be used against them in the future. Through alleviating tensions surrounding territorial disputes and gradually increasing military exchanges and transparency, China has tried to show that as it gets stronger it will be a force for peace. Although these achievements in the military realm are nowhere near as extensive as those in the economic and political realm, ASEAN leaders recognize that improving military relations is a lengthy process and welcome China's recent steps in the right direction.

Territorial Disputes

At the end of the Cold War, China had remaining territorial disputes with several of its Southeast Asian neighbors, but it has recently taken steps to resolve or at least reduce the tension associated with most of them. China had land disputes with Laos and Vietnam, as well as disputes with Vietnam over proper demarcation in the Beibu Gulf. In a dispute that began in the 1970s, but escalated after the Cold War, six governments—China, Taiwan, Vietnam, the Philippines, Malaysia, and Brunei—claim all or some of the Spratly Islands and the surrounding maritime area in the South China Sea. These areas are valuable due to their proximity to shipping lanes, the believed presence of oil and natural gas resources, and abundant fishing areas in these waters. In addition to these multilateral maritime disputes, China has bilateral disputes with Vietnam and the Philippines. China's steps toward solving these disputes are no doubt motivated by its own interests, such as stabilizing its periphery, but these actions have also signaled to other regional actors that China is willing to negotiate, compromise, and behave responsibly, and that it is not determined to seize every piece of territory in the region.

China has made great progress in working toward a resolution of territorial disputes with its southern neighbors, in many cases agreeing to divide disputed territory. China signed a border agreement with Laos in 1991 and with Vietnam in 1999. In 2000, China and Vietnam also signed an agreement regarding maritime boundaries and fishery cooperation in the Beibu Gulf. Although these arrangements are still being implemented, and many in Vietnam still worry about the possibility of China's future domination, these agreements illustrate the great progress in Sino-Vietnamese relations.[51]

China's military occupation of Mischief Reef in 1995 was the first time that China had challenged a member of ASEAN militarily over a territorial dispute—in this case, the Philippines. Although China made several conciliatory gestures after seizing the reef, it continued its pattern of "creeping assertiveness." In October 1998, China violated the spirit of the code of conduct it had signed with the Philippines by building new structures on Mischief Reef, including a ground satellite station, a structure that looked to be a potential helicopter pad, and what appeared to be gun emplacements.[52] In the late 1990s, however, China moderated its behavior toward the South China Sea, seizing no more territories and committing itself to multilateral discussions on establishing a code of conduct. Moreover, it has shown patience and restraint when other claimants have taken assertive action, such as when Malaysia moved to occupy Investigator Shoal and Erica Reef in May 1999, and when Vietnam raised the idea of letting tourists visit some areas of the South China Sea under its control.

After several years of discussions on a code of conduct, the claimants (except Taiwan) signed a Declaration on the Conduct of Parties in the South China Sea at the November 2002 APT summit in Phnom Penh. In the declaration, the parties pledged not to seize uninhabited islands or features.[53] It did not, however, prohibit building on territory that was already occupied, which displeased the Philippines. It also did not specifically mention the geographical scope of the declaration, which displeased Vietnam due to its separate dispute with China over the Paracel Islands. The biggest criticism of the declaration is that it is an informal, nonbinding document that places no real constraints on China's ability to take aggressive actions, if it chooses to do so.[54]

Even if this may seem like empty talk to some analysts in the West, all states involved in the disputes in the South China Sea—even Vietnam and the Philippines that feel the most threatened—are satisfied that such a declaration is a step in the right direction and will help build trust and goodwill that will create conditions more conducive to resolving these disputes. Perhaps building on the foundation of the declaration, the state-owned oil companies of China, Vietnam, and the Philippines agreed in March 2005 to jointly conduct seismic surveys and gather data on the potential for oil and gas reserves in the area.[55] In discussions on the implementation of the declaration, the parties are continuing to move toward signing a formal code of conduct.[56] China's more moderate policy is a positive development for the region and for Sino-

ASEAN relations, and even though there is no formal constraint contained in the declaration, China would incur reputation costs and image costs if it violated its spirit with a new assertive move.

China's military occupations of islands in 1974, 1988, and 1995 have all come at the expense of Vietnam or the Philippines, making these countries particularly wary of China's capabilities and intentions in maritime Southeast Asia. China's bilateral maritime disputes with Vietnam and the Philippines have not moved significantly toward resolution, but at least the tensions have subsided somewhat. Although China and Vietnam have agreed to the demarcation of the Beibu Gulf, Vietnam still claims sovereignty over the Paracel Islands, which China began occupying in the 1950s. Vietnam remains fearful that China may try to fortify these islands in the future and increase its power projection and ability to threaten Vietnam.[57] China also has a bilateral dispute with the Philippines over Scarborough Shoal, located 130 miles west of Luzon. Although Chinese naval vessels have occasionally been spotted in the area, the dispute mostly consists of incursions by Chinese fishermen who are forced to leave by the Philippine Navy.[58] These incursions continue, but there has never been any military escalation, even after mini-crises such as when a Chinese fishing vessel sank after colliding with a vessel of the Philippine Navy in May 1999, and after instances when Chinese fishermen have been imprisoned.[59] Although the Philippines and Vietnam appreciate the multilateral developments, they remain worried about whether or not a stronger China may try to enforce its claims in the South China Sea at their expense.[60]

Military Exchanges, Cooperation, and Transparency

China has demonstrated an increasing commitment to multilateral discussions of defense and security affairs by attending, hosting, and even proposing meetings and summits. The most important example is that China proposed the establishment of an ASEAN Security Policy Conference at the 2003 ARF summit and hosted its first meeting in November 2004.[61]

China has also used defense aid, military exchanges, and naval port calls to build confidence and reduce ASEAN's mistrust of China. China maintains the closest military relationship with Myanmar, which depends on China for most of its military aid and equipment, and has reportedly provided China with a listening post on the Indian Ocean.[62]

China has made an active effort in recent years to militarily engage the countries that remain the most wary of China. In addition to annual defense consultation talks that already exist with Thailand, China held its first round of annual talks with Vietnam and the Philippines (which included China proposing joint exercises) in 2005, and suggested the establishment of a similar mechanism to Indonesia. China's military assistance has also included Malaysia's purchase of anti-aircraft missiles in 2004, a $1.2 million donation of engineering equipment to the Philippines in 2005, and the signing of a memorandum of understanding on research and development in defense technology cooperation with Indonesia in 2005.[63]

China has also taken steps to increase military transparency in the hopes that providing such information may help to assuage fears of China's military modernization. In 1995, China issued a White Paper on Arms Control, and starting in 1998 has released a Defense White Paper every other year. Although critics charge that these documents contain vague generalizations and little specific information, each successive edition has contained increasingly specific information about China's military exchanges and joint exercises.[64]

China also appears increasingly willing to hold joint exercises with ASEAN. In 2001, China proposed that ARF members report on and send observers to multilateral joint military exercises. Since 2002, China has observed Cobra Gold, the joint military exercise involving the United States, Thailand, and Singapore. All ASEAN countries except Singapore observed China's "Iron Fist-2004," which was a mechanized infantry division exercise.[65] In 2002, China and ASEAN signed a declaration on cooperation against nontraditional security threats, and this may prove to be an area for enhanced cooperation and joint exercises.[66] In 2004, a Chinese senior colonel proposed anti-terrorism intelligence exchanges, joint naval exercises, and naval patrols.[67]

Again, Southeast Asian leaders seem pleased to see China moving in the right direction on military transparency and confidence-building measures, even if it is doing so slowly. How China might affect the overall balance of power in the region and apply its power over the long term, however, remain fundamental uncertainties for Southeast Asian nations.

Soft Power: The Subtle Tool of Influence

Although measuring soft power is very difficult, many have argued that China is at least beginning to show some signs of accumulating soft power in Southeast Asia, and such growing influence is leading to closer Sino-ASEAN relations.[68] For economically backward states with Communist or authoritarian political systems, such as Cambodia, Laos, Myanmar, and Vietnam, China's development path has become an object of study and emulation. According to one analyst, "Chinese culture, cuisine, calligraphy, cinema, curios, art, acupuncture, herbal medicine, and fashion fads have all emerged in regional (Southeast Asian) culture."[69] Chinese music, movies, and name brands have also become more popular throughout Southeast Asia, and China has signed formal agreements on cultural cooperation with several ASEAN states.[70]

Table 7.6

Students from Southeast Asia Studying in China

	No. of students in 2002	No. of students in 2003	No. of students in 2004	% change 2002–2004
Brunei	4	4	6	50
Cambodia	151	139	163	19
Indonesia	2,583	2,563	3,750	45
Laos	333	403	509	53
Malaysia	840	841	1,241	48
Myanmar	232	232	397	71
Philippines	638	602	1,375	115
Singapore	583	551	929	59
Thailand	1,737	1,554	2,371	36
Vietnam	2,336	3,487	4,382	88

Source: Zhonghua Renmin Gongheguo Waijiaobu (PRC Ministry of Foreign Affairs), *Zhongguo Waijiao* (China's diplomacy) (Beijing: World Affairs Press, 2003–2005).

Study of Mandarin Chinese in Southeast Asia has skyrocketed in recent years, and the Chinese government has offered assistance to improve the teaching of Mandarin in ASEAN countries.[71] Table 7.6 also illustrates the increasing appeal of studying abroad in China for Southeast Asian students. China has also started to attract more tourists from Southeast Asia, and has encouraged "roots travel," as a way to bring overseas Chinese back to the mainland. Although it is still too early to judge how much soft power China has actually accumulated, and whether it will be sustained over time, China's interactions and cooperation have increased its attractiveness in recent years.

Is China Becoming the Most Dominant Country in Southeast Asia?

Only looking at the recent advances in Sino-ASEAN relations may give one the impression that China is making significant gains at the expense of other East Asian actors, such as the United States and Japan. Some scholars and government officials, adopting a zero-sum view of the region, suggest that China's recent efforts to improve relations with Southeast Asia represent the reappearance of a competitive "great game," and further suggest that these improvements have damaged U.S. interests in the region.[72] In fact, both the United States and Japan remain positioned to wield considerable influence in Southeast Asia.

Moreover, there is not much evidence that China's recent initiatives have damaged U.S. interests in Southeast Asia. On the contrary, there is even some evidence to suggest that China's recent improvements in relations with Southeast Asia have helped protect certain U.S. regional interests. Building trust in the Sino-ASEAN relationship has reduced the chances of regional instability in general, and the Declaration on Conduct has reduced the threat of war in the South China Sea, which would threaten the sea lanes. China's economic initiatives will likely increase economic growth in Southeast Asian countries, which will help increase U.S.-ASEAN trade flows and the profitability of U.S. investments in the region. Moreover, improvements in Sino-ASEAN economic relations will help ASEAN countries become more politically stable, which will make it more difficult for terrorists to operate and more likely that democracy can take further hold in Southeast Asia. China's relationship with Southeast Asia has also forced China to join international and regional organizations that will give China a stake in the system, and hopefully moderate its behavior.

China's influence over continental Southeast Asia, including Thailand, is greater than its influence over maritime Southeast Asia. China, however, has not pushed any of these countries to give up things they care about or to side with China against the United States, so there have not been any real tests of how strong China's influence really is.[73] China's overall economic influence over Southeast Asia, although it has surely increased in recent years, remains minimal. According to 2005 IMF data, only Myanmar and Vietnam conduct more than 10 percent of their total trade with China, and only Myanmar and Laos conduct more of their total trade with China than with the United States. China's investment and overseas development assistance (ODA) are still very small. Many of the agreements and declarations that have won China goodwill in Southeast Asia have yet to be implemented.

Even after a period of Japanese economic weakness and decline, and a period when the United States has been perceived as uninterested in Southeast Asia or too focused on fighting terrorism, the United States and Japan still remain in a relatively solid position to wield influence in Southeast Asia. They will continue to serve as important sources of technology for ASEAN countries and important markets for Southeast Asia's exports of manufactured products. Although Sino-ASEAN bilateral growth has exploded, reaching $105.9 billion in 2004, Japan-ASEAN total bilateral trade ($127 billion in 2004) and U.S.-ASEAN trade ($136 billion in 2004) still surpass China's regional trade.[74] Sino-ASEAN bilateral trade is likely to bypass trade with the United States and Japan in the next three to five years. These trade data, however, understate the continued importance of the U.S. market to both China and Southeast Asia, as many exports from

Southeast Asia to China are processed into final products for ultimate export to the United States.

China's FDI to Southeast Asia failed to reach $1 billion in 2001, while U.S. FDI was $3.4 billion and Japanese FDI was $4.3 billion.[75] In 2003, the total U.S. FDI position in the region was $88 billion, and many suggest this underestimates the total value of U.S. investment.[76] Japan gave considerable aid to Southeast Asia during the financial crisis and, by one estimate, has provided over $23 billion in ODA to the region over the last three decades.[77] In terms of trade, and even more in terms of investment, the United States and Japan remain in a very solid position to wield influence in Southeast Asia.

The United States and Japan have taken action to improve relations with Southeast Asia, although probably not as quickly as China. The United States has only signed an FTA with Singapore, but is negotiating several other bilateral FTAs. Japan, although hobbled by its powerful protectionist agriculture lobby, has agreed to implement several bilateral FTAs and is in the process of negotiating a regional FTA. The United States is trying to correct the perception that it does not care about the region through recent initiatives such as the *Joint Vision Statement on the ASEAN-U.S. Enhanced Partnership*, issued in November 2005, and discussions of an annual summit on the side of APEC between the United States and the seven ASEAN countries that belong to APEC.[78] Perhaps of greatest significance, the United States strengthened its military relations with alliance partners and friends throughout maritime Southeast Asia over concerns of terrorism and uncertainty about China's intentions. The United States has continued to hold several bilateral and multilateral joint exercises, the Philippines and Thailand have both been designated as "major non-NATO allies," the United States signed a Strategic Framework Agreement with Singapore in 2005, and, after the tsunami, military relations with Indonesia have significantly improved.[79]

Although China has improved its relations with ASEAN and increased its regional influence, it has (for the most part) not directly challenged U.S. regional interests. Although its support for initiatives such as the APT and ACFTA have helped to develop pan-Asianism, China has not pushed for a closed regional economic system that would leave the United States on the outside. Given the region's huge variation in political systems and levels of economic development; deep residual mistrust; and the importance of the United States as a source of investment and markets, for exports (including for China), any attempts in the near future to exclude the United States from political, economic, or security affairs will likely garner little support.

There is also little evidence to suggest that China's improved relations with Southeast Asia have damaged United States' alliances and military presence in the region. Even though many in China remain suspicious about United States' intentions, the Chinese government has recently moderated its criticisms of U.S. alliances and even referred to the U.S. presence in the Asia-Pacific as a "stabilizing factor."[80] Given lingering uncertainty about the implications of China's rise, any future attempt by China to pressure ASEAN countries to end their relationships with the United States will prove not only unsuccessful but counterproductive, as it would make regional states more suspicious about China's intentions.

Improved Sino-ASEAN relations have certainly contributed to Southeast Asia's open rebukes of Taiwanese independence and some of President Chen Shui-bian's statements and initiatives, which some consider to be a challenge to U.S. regional interests. Perceptions that Taiwan has become the greater risk to the status quo and regional stability, however, are probably a more important cause of this shift. These rebukes of Taiwanese independence by officials in Southeast Asian countries, most notably Lee Hsien Loong's statements, have usually included declarations of nonsupport or opposition to Taiwanese independence, or assertions that Taiwan's declaration of independence is not in the national interest of the Southeast Asian countries. It is difficult to characterize these shifts in position on Taiwan as *against* U.S. national interests because the new position is virtually identical to the U.S. government's position on Taiwan.

The one area in which China's growing influence has undeniably challenged U.S. interests is in its support for nondemocratic regimes. China's support for authoritarian regimes in Cambodia and Laos—and in particular its influence over Myanmar, which has become very dependent on Chinese military aid, trade, investment, and political support for sustenance—run counter to U.S. interests to promote human rights and democracy in Asia. Chinese Foreign Minister Li Zhaoxing's decision to skip the 2005 ARF summit in favor of a formal visit to Myanmar was a demonstration of how deep China's commitment to solidify its economic, political, and military position in Myanmar is, despite the presence of a brutal regime that has embarrassed ASEAN members. ASEAN's concern about China's growing influence in Myanmar, in fact, is an area of concern in the region that has raised questions about China's ultimate role and underlying intentions in regional affairs.

Finally, Chinese dominance of Southeast Asia will be mitigated by the strategy of balancing or "counter-dominance" that most ASEAN countries have adopted. In the post-Cold War world, most ASEAN states have tried to maintain cordial relations with all of the major powers, allowing them to benefit from all and not become too dependent on any one state.[81] Simply put, ASEAN states do not want to choose sides between the United States or China, but rather hope to maximize benefits by maintaining good relations with both. Any Chinese attempt to form an anti-U.S. coalition or American attempt to form an anti-China coalition is destined to fail. Southeast Asian nations have shown in the past that when they have felt threatened, such as in the mid-1990s when China was behaving aggressively in the South China Sea and in the Taiwan Strait, they have taken steps to move in the direction of balancing against that threat by improving relations with other outside powers.

The results of the East Asia Summit (EAS), held in Kuala Lumpur in December 2005, also showed that worries about a closed regional organization that excludes the United States but also allows China to play a formal leadership role were premature. Once the EAS was expanded to include India, Australia, and New Zealand, countries that seemed to prefer a closed organization, such as China and Malaysia, seemed less interested in it.

The parties compromised, and according to the APT and EAS summit statement, decided that the "ASEAN Plus Three process will continue to be the main vehicle in achieving that goal" (East Asian cooperation). EAS, on the other hand, would be "an open, inclusive, transparent and outward-looking forum." An important phrase which appeared in the statements from both summits, "with ASEAN as the driving force," reassured the ASEAN states that they remained in the leadership position.[82] China did not get its best outcome from these recent summits, but the compromise allowed it to avoid potential friction in its relations with Southeast Asia and the United States. Although China's attempts to portray its position as having always supported ASEAN as the driver of East Asian cooperation is a little disingenuous, given its push to host the EAS and hopes that the EAS would become the driver of regionalism, the ASEAN countries were very pleased by the Chinese official declaration of support for ASEAN's leading role.[83] China also learned that if it were to push for a regional organization that does not include the United States, but allows it to play a leadership role, it would not garner much support in the future. Although creating such a grouping may still be China's ultimate goal, the realization that it is unattainable in the near future will help stabilize China's relations with Southeast Asia and the United States.

China's influence in Southeast Asia has increased in recent years, but there is no need for Washington to overreact. The lackluster U.S. response to the regional financial crisis during the Clinton administration and the focus on anti-terrorism during the Bush administration helped provide China with an excellent opportunity to improve relations with the ASEAN states by addressing the needs of regional states. Rather than turning this into a competition over influence in Southeast Asia and trying to convince the ASEAN states that China's behavior is a threat to them, U.S. policy can better prevent China's domination of Southeast Asia by engaging the region on issues that are important to it, such as trade, finance, and internal domestic instability. Deputy

Secretary of State Robert Zoellick's ten-day visit to Singapore, Thailand, the Philippines, Indonesia, Malaysia, and Vietnam in May 2005 seemed to demonstrate just this type of comprehensive approach to relations with Southeast Asia. Zoellick genuinely listened to Southeast Asia's problems, focused on the need to consult friends in Southeast Asia rather than to make demands on them, devoted considerable attention to developing economic ties, and repeated that the United States was neither trying to contain nor compete with China.[84] The consistency of this moderate approach, however, was called into question in the minds of many in Southeast Asia when Secretary of State Condoleezza Rice skipped the July 2005 ARF summit, citing "other vital travel."[85] Hopefully, initiatives such as holding a summit with ASEAN countries on the side of APEC will come to fruition and will lead to a consistent U.S. policy of comprehensively engaging ASEAN in the future.

Maintaining the Momentum in China's Policy Toward Southeast Asia

Looking to the future, some worry that if Sino-ASEAN relations continue to improve at the rate they have in recent years, China's regional influence will soon surpass that of the United States. These concerns, however, overlook the future obstacles that China will likely need to overcome and trade-offs that it will need to manage in its future relations with Southeast Asia. This is not to say that these obstacles are insurmountable, but they are likely to complicate the future development of Sino-ASEAN relations.

What If ASEAN States Are Unable to Compete and China's Future Growth Doesn't Trickle Down?

Initiatives like the ACFTA and Early Harvest Program (EHP) were designed to give ASEAN states a preferential opportunity to penetrate the China market. As mentioned earlier, however, if goods from these countries are unable to compete in China's domestic market, the goodwill that China accumulated through the announcement of these initiatives will likely disappear very quickly. Thailand's and the Philippines' experiences with the EHP, in which they are flooded by cheaper Chinese agricultural goods and are having trouble competing in the Chinese market, may prove to be a harbinger of things to come.[86] If Southeast Asia is unable to compete in the Chinese or global markets, the chances of substantial Chinese investment flowing into ASEAN in the future will also be pretty low. If China's growth does not spread to the region, there is likely to be increased resentment and ASEAN leaders will be more likely to challenge China on smaller issues, leading to a more conflicting relationship.

How Will China Address Worries About a Return to a Colonial Economic Relationship?

China hopes to secure access to raw materials in Southeast Asia such as oil, natural gas, rubber, and tin, and has focused much of its FDI on projects devoted to the exploitation of natural resources. If ASEAN countries do not work to increase productivity and develop market niches, especially as China begins to develop more high-tech products, competition from China may leave the manufacturing capacity of ASEAN states in disarray and force them to return to the colonial situation in which they relied on exports of raw materials. A relationship which forces ASEAN countries into less profitable sectors and into a subordinate political and economic position would likely produce some sort of backlash. Some analysts warn that the ACFTA may end up creating this type of colonial relationship.[87]

How Will China Balance Its Growing Oil and Energy Demands and Its Need to Maintain Cordial Relations with Its Neighbors?

China's need for oil, gas, and other natural resources, as well as its need to find greater supplies of power for the poorer western provinces, may lead to tension and conflict in its future relations with Southeast Asia. China has started to implement an extensive dam-building project on its part of the Mekong River to generate hydro-electric power, with two dams already completed and several more to be built in the future. Several studies have shown that these dams have already led to disastrous drops in the amount of fish caught downstream in Indochina, and if the dam projects continue they may cause irreparable harm to the local economies. These downstream countries rely on fish, not only as one of the foundations of the local economy, but as the most important source of protein for their population. In large part because these countries are hoping that maintaining good relations with China will yield economic windfalls in the future, open challenges to China on the disastrous effects of the dams have been limited. If China's growth does not spill over into these countries, or if the fishing industry is completely destroyed, disputes like the damming of the Mekong may become a point of irritation in the Sino-ASEAN relationship.[88]

China has also invested heavily in a plan to clear land in Kalimantan, in the Indonesian part of Borneo, to make way for a palm oil plantation. The cost of more access to palm oil—used to make several daily household products, as well as biofuel—will include the destruction of one of the world's remaining rainforests and its rich biodiversity, and increase tension with Singapore and

Malaysia from the environmental hazards of the smoke from the resulting forest fires.[89] The oil and natural resource-rich South China Sea may also become an area where China's demands for future resources may force them to take actions at the expense of the Southeast Asian countries.

Will China's Continued Military Modernization and Threats Aimed at Taiwan Undermine Its Charm Offensive?

Southeast Asia benefits greatly from its trade with and investment from Taiwan. ASEAN countries therefore recognize that a war over Taiwan that could destabilize the region is not in their best interests. While there has been a detectable shift in sympathy away from Taiwan over the last five years due to concerns about the perceived provocations of Taiwan President Chen, ASEAN countries also understand that the power projection capabilities that the PLA is developing to take back Taiwan are the same type of capabilities that the PLA would need to threaten maritime Southeast Asia. When China tries to intimidate Taiwan through threats and military exercises, such as in 1995 and 1996, ASEAN countries worry about China's future intentions in the region. China's future military modernization, especially if it is combined with more threats against Taiwan, are likely to make it more difficult to believe that China is a country with nothing but peaceful intentions in the region.

Will Worries over Taiwan Force China to Take Actions That Are Against the Interests of ASEAN States?

If concerns about Taiwan lead China to pressure Southeast Asian states to do things against their own interests, this could put a strain on Sino-ASEAN relations. There have been several examples of this situation over the last ten years. Due to disagreements over the involvement of Taiwan, China waited three years to join CSCAP—the Track II version of the ARF—which was devoted to building confidence and trust in security affairs. China boycotted the 2004 Shangri La security conference due to the presence of a low-level Taiwanese delegate. China's public rebuke of Singapore's Prime Minister-designate Lee Hsien Loong after his visit to Taiwan in June 2004 shows that concerns over Taiwan can also affect bilateral relations.[90] Future proposals of a Taiwan-ASEAN FTA or attempts to include Taiwan in Asian institutions are two conceivable situations in which China would be put in a difficult situation, and might be forced to sacrifice its relations with Southeast Asia for the sake of limiting Taiwan's international space and prestige.

Will China Respond to U.S. and Japanese Initiatives to Improve Their Relations with Southeast Asia by Turning It into a Competition?

As the United States and Japan engage with Southeast Asia, some voices in China do not see this as a win-win situation and are attacking U.S. and Japanese motives. Many Chinese observers, for example, have suggested that the U.S. motive behind improving relations with Vietnam is to contain (*ezhi*) China.[91] Some analysts suggest that U.S. counter-terrorism initiatives in Southeast Asia and assistance after the December 2004 tsunami were merely excuses for the U.S. military to return to Southeast Asia or conduct realistic training exercises. Some analysts criticized Japanese assistance after the tsunami as motivated solely by its desire to become a permanent member of the UN Security Council.[92] Although some suspicion over motives is natural, if China's response to future initiatives by the United States, Japan, and other regional powers in Southeast Asia is to consider them as part of a competition, China may end up damaging Sino-ASEAN relations.

Conclusion

A Chinese analyst claims that, as a result of China's focused efforts to improve its relations with Southeast Asia, "the ASEAN countries' worries and misgivings about China have been fundamentally eliminated and dispelled (*jiben xiaochu*), and have been replaced with trust, admiration, and respect."[93] Continuing concerns among ASEAN states about the possibility of a stronger China trying to become the regional hegemon suggest that China's actions have not fundamentally eliminated all worries, but China's active policy to address Southeast Asia's fears about the future impact of a stronger China have made these states more optimistic and hopeful that a more powerful China will be a force for peace, stability, and prosperity in the short to medium term. China's ability to reduce mistrust and suspicion through its policy toward ASEAN has helped China stabilize its periphery, reduce the chances of the formation of an anti-China coalition, and create constructive external conditions for China to focus on its economic modernization—and each are critical goals of the Beijing regime. China's improved relations have also helped China gain access to vital raw materials and investment from Southeast Asia, as well as increased bilateral trade flows. If China's aim is to become the dominant player in Southeast Asia, it will have to overcome future potential obstacles in Sino-ASEAN relations, but China's proactive policies have clearly improved its position in Southeast Asia.

Regional responses to the devastating December 2004 tsunami showed that China is trying to do more and more to engage Southeast Asia, but the responses of Japan and the United States demonstrated the relative weakness of China's material capabilities. In fact, the United States and Japan remain well-positioned to wield considerably greater influence in the region. However, both the United States and Japan should not be complacent but should work harder to listen to the concerns of ASEAN countries and focus energy on trying to address their concerns while demonstrating commitment to the values and ethics of the region, including multilateral approaches to solving regional problems.

Even though the United States and Japan remain well-positioned in Southeast Asia, China's increasing influence in the region is a challenge that the United States will need to confront. U.S. analysts should recognize that adopting an openly competitive, zero-sum view of China's relations with Southeast Asia is not in the United States' interests. Trying to force the Southeast Asian countries into an anti-China coalition, especially doing so before China startes behaving aggressively, would most likely backfire and antagonize the Southeast Asian nations. Instead, U.S. leaders should recognize that much of China's improved relations comes from China's ability to address a wide range of issues that worry ASEAN leaders. Rather than focusing on the military aspects of anti-terrorism, the United States should adopt a more comprehensive approach and spend more effort on economic and political issues in addition to the military aspects of anti-terrorism. Depending on how China, the United States, Japan, and India respond to each other's initiatives to improve relations with the ASEAN countries, future international relations in Southeast Asia could turn into a truly win-win situation with all countries improving relations with ASEAN, or they could devolve into cutthroat competition.

Notes

The author is indebted to Michael Chambers, Bernard Cole, Richard Cronin, Matthew Ferchen, Taylor Fravel, Paul Godwin, Eric Heginbotham, Li Nan, Matthew Oresman Bronson Percival, Ren Xiao, Sheng Lijun, Jenny Scala, Ian Storey, Robert Sutter, Tang Shiping, Allen Whiting, and Michael Wills for comments and suggestions. He is especially grateful to Tom Christensen and Derek Mitchell for providing particularly detailed feedback. A more thorough version of this chapter appeared as "Heading toward a Win-Win Future? Recent Developments in China's Policy toward Southeast Asia," *Asian Security,* vol. 2, no. 1 (2006): 24–57.

1. Statement by Christopher R. Hill, Assistant Secretary of State for East Asian and Pacific Affairs, "Emergence of China in the Asia-Pacific: Economic and Security Consequences for the United States," Senate Foreign Relations Committee, June 7, 2005. Available at foreign.senate.gov/testimony/2005/HillTestimony050607.pdf.

2. U.S. analysts examining China's Asia policy have concluded that its greatest successes have been with ASEAN and South Korea. For examples, see Robert Sutter,

"China's Recent Approach to Asia: Seeking Long-term Gains," *NBR Analysis*, vol. 13, no. 1 (March 2002): 13–38; and David Shambaugh, "China Engages Asia: Reshaping the Regional Order," *International Security*, vol. 29, no. 3 (Winter 2004/5): 64–99. China's Deputy Foreign Minister Wu Dawei recently declared, "China-ASEAN Relations Are at its Best in History." See "Foreign Ministry Official Gives a Briefing on the Background of President Hu Jintao's Upcoming Visit to Southeast Asian Countries," (April 18, 2005). Available at www.fmprc.gov.cn/eng/xwfw/wgjzxwzx/ipccfw/t192598.htm. A recent comprehensive Chinese book on Sino-ASEAN relations repeatedly makes reference to relations rising to a new level and achieving unprecedented levels of cooperation. See Cao Yunhua and Tang Chong, *Xin Zhongguo-Dongmeng Guanxi Lun* (New Sino-ASEAN relations) (Beijing: World Affairs Press, 2005). For a recent declaration through ASEAN of the closeness of the relationship, see "Deepening ASEAN-China Strategic Partnership," Chairman's Statement of the 8th ASEAN + China Summit (November 29, 2004). Available at www.aseansec.org/16749.htm.

3. For a recent analysis of related issues, see Bruce Vaughn, "China-Southeast Asia Relations: Trends, Issues, and Implications for the United States," *Congressional Research Service Report for Congress* (December 7, 2004); Alice Ba, "China and ASEAN: Renavigating Relations for a 21st-Century Asia," *Asian Survey*, vol. 43, no. 4 (2003): 622–647; and Denny Roy, "China and Southeast Asia: ASEAN Makes the Best of the Inevitable," *Asia-Pacific Center for Security Studies*, vol. 1, no. 4 (November 2002). For recent analysis of China's Asia policy, see David Shambaugh, "China Engages Asia: Reshaping the Regional Order," *International Security*, vol. 29, no. 3 (Winter 2004/5): 64–99. For the most thorough recent Chinese analysis, see Cao Yunhua and Tang Chong, *Xin Zhongguo-Dongmeng Guanxi Lun* (New Sino-ASEAN relations) (Beijing: World Affairs Press, 2005).

4. See "Full Text of Jiang Zemin's Report at 16th Party Congress," Xinhua (November 17, 2002). Available at http://www.china.org.cn/english/features/49007.htm. Another translation that is often used for this phrase is "treating neighbors with goodwill and coexisting with neighbors as partners." See "Speech at ASEAN Business and Investment Summit by Mr. Wen Jiabao, Premier of the State Council of the People's Republic of China" (November 10, 2003). Available at www.fmprc.gov.cn/eng/topics/zgcydyhz/dqc/t27711.htm. See "Wen Spells out China's 'Friendly Elephant' Role," *Straits Times* (March 15, 2004). For the complete text of the press conference, see "Premier Wen Jiabao's Press Conference at the Conclusion of the Second Session of the 10th National People's Congress" (March 20, 2004). Available at www.chinaembassycanada.org/eng/xwdt/t80426.htm.

5. For more on China's support for Communist revolutions, see Jay Taylor, *China and Southeast Asia: Peking's Relations with Revolutionary Movements* (New York: Praeger, 1976); and Peter Van Ness, *Revolution and Chinese Foreign Policy: Peking's Support for Wars of National Liberation* (Berkeley, CA: University of California Press, 1971). Some Chinese scholars have recognized the link between these Cold War policies and ASEAN's fears about the intentions of a rising China in the 1990s and beyond. See *Weilai 10–15 nian Zhongguo zai Yatai Diqu Mianlin de Guoji Huanjing* (China's security environment in the next ten to fifteen years), ed. Zhang Yunling (Beijing: Chinese Academy of Social Sciences Press, 2003), 289–319; *Zhongguo Zhoubian Anquan Huanjing Toushi* (Perspectives on China's peripheral security environment), ed. Yang Chengxu (Beijing: China Youth Press, 2003), 334–342.

6. When ASEAN was initially established in 1967, its membership included Indonesia, Malaysia, Philippines, Singapore, and Thailand. Brunei joined in 1984,

Vietnam joined in 1995, Myanmar and Laos joined in 1997, and Cambodia joined in 1999. Although recognizing that ASEAN is technically an international organization, this chapter sometimes uses the terms Southeast Asia and ASEAN interchangeably.

7. See Allen S. Whiting, "ASEAN Eyes China: The Security Dimension," *Asian Survey,* vol. 37, no. 4 (April 1997): 299–322; and Derek Da Cunha, "Southeast Asian Perceptions of China's Future Security Role in Its 'Backyard,'" in *In China's Shadow: Regional Perspectives on Chinese Foreign Policy and Military Development,* ed. Jonathan D. Pollack and Richard H. Yang (Santa Monica, CA: RAND, 1998), 115–126. For discussions by Chinese scholars of how these aggressive moves were backfiring, see Zhang Xishen, "Zhongguo tong Dongmeng de Mulin Huxin Huoban Guanxi" (Sino-ASEAN partnership relations of good-neighborliness and mutual trust), *Dangdai Yatai* (Contemporary Asia-Pacific), no. 2 (1999): 25–29; and Wang Jisi, "China's Changing Role in Asia," Atlantic Council of the United States (January 2004): 9. Although willingness to move toward joining an anti-China coalition was far from uniform throughout the region, the above works detail several examples of such a tendency, including Indonesia's military agreement with Australia signed in 1995, improved U.S. military-to-military cooperation with regional countries such as the Philippines and Singapore, and Vietnam's decision to join ASEAN.

8. For some examples, see "The Growth and Limits of China's Reach in Southeast Asia," *Nation* (October 4, 2004); and S. Pushpanathan, "Building an ASEAN-China Strategic Partnership," *Jakarta Post* (July 1, 2004). For an analysis of Southeast Asia's response to the rise of China, see Amitav Acharya, "Seeking Security in the Dragon's Shadow: China and Southeast Asia in the Emerging Asian Order," *IDSS Working Paper* (March 2003).

9. Most Chinese-language analyses of Chinese foreign policy in general, as well as Sino-ASEAN relations, begin with a discussion of the number of countries on China's periphery. For examples, see Zhang Yunling, chief ed., *Weilai 10–15 nian Zhongguo zai Yatai Diqu Mianlin de Guoji Huanjing* (hereafter *Weilai 10–15 nian);* Hou Songling and Chi Diantang, "Dongnanya yu Zhongya: Zhongguo zai Xin Shiji de Diyuan Zhanlue Xuanze" (Southeast Asia and Central Asia: China's regional strategic choices in the new century), *Dangdai Yatai* (Contemporary Asia-Pacific), (April 2003): 9–15; Chen Fengjun, "Jiaqiang Zhongguo yu Dongmeng Hezuo de Zhanlue Yiyi" (Augmenting the strategic meaning of Sino-ASEAN cooperation), *Guoji Zhengzhi Yanjiu* (Studies of international politics), no. 1 (February 2004): 24–28; and Lu Jianren, "Shijizhijiao: Zhongguo dui Dongmeng de Waijiao Zhanlue" (At the turn of the century: China's diplomatic strategy toward ASEAN), *Taipingyang Xuebao* (Pacific journal), no. 1 (1998): 42–47. For English-language analysis, see Zhao Suisheng, "China's Periphery Policy and its Asian Neighbors," *Security Dialogue,* vol. 30, no. 3 (1999): 335–346.

10. According to one Chinese analyst, China was invaded seven times in the last century by sea from the south and the east. See Ji Guoxing, "China versus South China Sea Security," *Security Dialogue,* vol. 29, no. 1 (1998): 102.

11. See Chen Fengjun, "Jiaqiang Zhongguo yu Dongmeng Hezuo de Zhanlue Yiyi" (Augmenting the strategic meaning of Sino-ASEAN cooperation), 25; and Hou Songling and Chi Diantang, "Dongnanya yu Zhongya: Zhongguo zai Xin Shiji de Diyuan Zhanlue Xuanze" (Southeast Asia and Central Asia: China's regional strategic choices in the new century), 11. According to one analyst, "Right now the United States does not have troops that are forward deployed for a long period of time in Southeast Asia . . . as long as China and Southeast Asia develop close relations, Southeast Asia

will not have a reason to invite the United States to come in." See Zhang Yunling, *Weilai 10-15 nian,* 309.

12. See Huang Renwei, *Zhongguo Jueqi de Shijian he Kongjian* (Time and space for China's rise) (Shanghai: Shanghai Academy of Social Sciences Press, 2002); Guo Xuetang, "Zhua Zhanlue Jiyu, Bi Zhanlue Fengxian" (Grasp strategic opportunity, avoid strategic hazards), *Huanqiu Shibao* (Global times) (February 21, 2003).

13. See Tang Shiping and Zhang Yunling, "Zhongguo de Diqu Zhanlue" (China's regional strategy), *Shijie Jingji yu Zhengzhi* (World economics and politics), no. 6 (2004): 8–13.

14. See Guo Xuetang, "Zhua Zhanlue Jiyu, Bi Zhanlue Fengxian" (Grasp strategic opportunity, avoid strategic risks), *Huanqiu Shibao* (Global times) (February 21, 2003). For a more detailed analysis of Chinese foreign policy that repeats this metaphor, see Guo Xuetang, "Xin Shiji Zhongguo duiwai Zhanlue de Beijing ji Sikao: yi zhong Lilun yu Xianshi Fenxi" (Background and thought on China's foreign strategy in the new century), *Tongji Daxue Xuebao: Shehui Kexue Ban* (Tongji University journal: Social sciences edition), vol. 14, no. 4 (August 2003): 10–15. The use of the metaphor of building a ring is particularly important because one of China's greatest fears is that other powers will build a ring of encirclement (*baowei quan*) on its periphery.

15. As will be discussed later in this chapter, and is discussed in other chapters in this volume, such a strategy of improving relations with neighbors may also end up adding new strains to Sino-U.S. relations if the United States interprets such activities as coming at the expense of its interests or its control in the region.

16. See Pang Zhongying, "Zhongguo de Yazhou Zhanlue: Linghuo de Duobianzhuyi" (China's Asia strategy: Flexible multilateralism), *Shijie Jingji yu Zhengzhi* (World economics and politics), no. 10 (2001): 33.

17. See Michael Richardson, "Japan's Lack of Leadership Pushes ASEAN toward Cooperation with China," *International Herald Tribune* (April 17, 1998); and Robert G. Lees, "If Japan Won't Help Rescue Asia, China Might," *International Herald Tribune* (February 21, 1998).

18. For a discussion of China's unselfish motives, see Yang Qing, "Zhongguo Heping Jueqi yu Zhongguo-Dongmeng Zhijian de Guanxi" (China's Peaceful Rise and Sino-ASEAN Relations) *Zhonggong Zhangyang Dangxiao Xuebao* (Journal of the CCP's Central Party School), vol. 8, no. 1 (February 2004), 125. Chinese analysts often cite China's behavior during the Asian financial crisis as an example of how China puts other countries' interests ahead of its own and how it has become a responsible power. Author's interviews, Beijing and Shanghai (Spring–Summer 2004). For a thorough discussion of China's motives, see Hongying Wang, "China's Exchange Rate Policy in the Aftermath of the Asian Financial Crisis," in *Monetary Orders: Ambiguous Economics, Ubiquitous Politics,* ed. Jonathan Kirshner (Ithaca, NY: Cornell University Press, 2003), 153–171.

19. See Richardson, "Japan's Lack of Leadership Pushes ASEAN Toward Cooperation with China."

20. ASEAN Plus Three includes the ten members of ASEAN, China, Japan, and South Korea.

21. See Jennifer Amyx, "What Motivates Regional Financial Cooperation in East Asia Today?" *Analysis from the East-West Center*, no. 76 (February 2005).

22. See Miyako Takebe and Monica Houson-Waesch, "ASEAN + 3 Agrees to Expand Currency-Swap Pacts," *Asian Wall Street Journal* (May 5, 2005).

23. For an analysis of China's evolving position on the Asian Monetary Fund, see

Yu Yongding, "On East Asian Monetary Cooperation," *Working Paper Series No. 2,* Chinese Academy of Social Sciences, Research Center for International Finance (August 2001). Interestingly, China's opposition to the fund is rarely mentioned in most ASEAN discussions about the Asian financial crisis, thus preventing damage to China's image.

24. See Michael Richardson, "China's Growth Weighs on Neighboring Countries," *International Herald Tribune* (August 18, 2001); and "ASEAN Notes Both Threats and Opportunities in China's Entry into WTO" (November 26, 1999), available at www.aseansec.org/11949.htm.

25. In Japan's relatively open textile market, for instance, China's share of the cotton knit apparel market has increased from 47.3 percent in 1996 to 77.3 percent in 2001, and its share of man-made fiber knit garments has increased from 59.1 percent to 80.4 percent over the same period. See Mari Pangestu, "China's Economic Rise and the Responses of ASEAN," in *The Rise of China and a Changing East Asian Order,* ed. Kokubun Ryosei and Wang Jisi (Tokyo: Japan Center for International Exchange, 2004), 241–263. See also Edmund L. Andrews, "Bush Administration Will Ask China to Agree to Broad Limits on Clothing Exports," *New York Times* (August 2, 2005).

26. For some examples of Chinese scholars openly discussing these worries, see Chen Fengjun, "Jiaqiang Zhongguo yu Dongmeng Hezuo de Zhanlue Yiyi" (Augmenting the strategic meaning of Sino-ASEAN cooperation), 24; Zhang Yunling, *Weilai 10-15 nian,* 302; and Cao Yunhua, "Dongmeng yu Daguo Guanxi Pingxi" (Analysis of ASEAN-Great Power relations), *Guoji Zhengzhi Yanjiu* (Studies of international politics), no. 2 (May 2003): 128.

27. See "Hu Calls for Closer China-ASEAN Economic Cooperation," *People's Daily* (April 26, 2005).

28. For the full report of the expert group, see ASEAN-China Expert Group on Economic Cooperation, "Forging Closer ASEAN-China Economic Relations in the 21st Century" (October 2001), available at www.aseansec.org/5883. For the complete text of the agreement to establish the FTA, see "Framework Agreement on Comprehensive Economic Co-operation Between the Association of Southeast Asian Nations and the People's Republic of China" (November 4, 2002), available at www.aseansec.org/13196.htm.

29. See "China-ASEAN Ties Dynamic and Concrete," *China Daily* (November 30, 2004).

30. For recent analysis of China's motivations and ACFTA, see Joseph Yu-shek Cheng, "The ASEAN-China Free Trade Area: Genesis and Implications," *Australian Journal of International Affairs,* vol. 58, no. 2 (June 2004): 257–277; John Wong and Sarah Chan, "China-ASEAN Free Trade Agreement: Shaping Future Economic Relations," *Asian Survey,* vol. 43, no. 3 (May/June 2003): 507–526; and Sheng Lijun "China-ASEAN Free Trade Area: Origins, Developments and Strategic Motivations," *ISEAS Working Paper, International Politics and Security Issues,* no. 1 (2003). For a Chinese article that not only stresses China's concessions in reaching the agreement, but the difficult road ahead in its implementation, see Zhang Yunling, "Zhongguo-Dongmeng Ziyou Maoyiqu de Jiyu he Tiaozhan" (Opportunities and challenges of the China-ASEAN free trade area), *Yatai Jingji* (Asia-Pacific economic review), no. 3 (2003): 2–4.

31. See Li Qingsi, "Zhongguo yu Dongmeng Guanxi" (Sino-ASEAN Relations), *Guoji Luntan* (International forum), vol. 6, no. 2 (March 2004), 31. However, these

early harvest agreements exclude rice and palm oil, two important exports for ASEAN. See Wong and Chan, "China-ASEAN Free Trade Agreement," 511.

32. See Shanghai Institute of International Studies analyst Ren Xiao's comments in Peter Kammerer, "Smiles All Around, but Suspicion of China Won't Go Away," *South China Morning Post* (December 1, 2004).

33. See UNCTAD, *World Investment Report 2004,* 298.

34. See Jiang Guocheng, "China Sets up Regional Poverty Reduction Fund Through ADB," Xinhua (March 25, 2005).

35. For more details and analysis, see Michael A. Glosny, "Meeting the Development Challenge in the 21st Century: American and Chinese Perspectives on Foreign Aid," *China Policy Series,* National Committee on U.S.-China Relations (2006).

36. See "Mekong Becomes Crucial Trade Channel for Neighboring Asian Countries," Xinhua (January 26, 2004).

37. Denny Roy, "China and Southeast Asia: ASEAN Makes the Best of the Inevitable," *Asia-Pacific Center for Security Studies*, vol. 1, no. 4 (November 2002), 3. Table 7.4 seems to show a large decline in investment flowing into mainland China from 2003 to 2004. During the same period, however, investment into Hong Kong increased from $13.6 billion to $34.0 billion. See UNCTAD, *World Investment Report, 2005.*

38. See Zhang Yunling, *Weilai 10–15 nian,* 311.

39. See Michael R. Vatikiotis, "Catching the Dragon's Tail: China and Southeast Asia in the 21st Century," *Contemporary Southeast Asia,* vol. 25, no. 1 (April 2003): 65–78.

40. See "Trade with ASEAN Expected to Boom," *China Daily* (October 8, 2003); and Jane Perlez, "China Promises More Investment in Southeast Asia," *New York Times* (October 8, 2003).

41. UNCTAD, *World Investment Report, 2004.* China has also only sent 10–20 percent of its total FDI to Southeast Asia. See John Wong and Sarah Chan, "China's Outward Direct Investment: Expanding Worldwide," *China: An International Journal,* vol. 1, no. 2 (September 2003): 273–301.

42. See Clarissa Oon, "Beijing Wants Jakarta as Strategic Partner," *Straits Times* (July 27, 2005).

43. See Edward J. Lincoln, *East Asian Economic Regionalism* (Washington, D.C.: Brookings, 2004), 88.

44. For discussions of China's fears of joining multilateral institutions and its gradual shift toward acceptance, see Alastair Iain Johnston and Paul M. Evans, "China's Engagement with Multilateral Security Institutions," in *Engaging China: The Management of an Emerging Power,* ed. Alastair Iain Johnston and Robert S. Ross (London: Routledge, 1999), 235–272; and Jing-dong Yuan, "Regional Institutions and Cooperative Security: Chinese Approaches and Policies," *Korean Journal of Defense Analysis,* vol. 13, no. 1 (Autumn 2001): 263–294.

45. It is difficult to explain China's decision to join the ARF without taking into account its desire to avoid negative image costs. See Alastair Iain Johnston, "The Myth of the ASEAN Way? Explaining the Evolution of the ASEAN Regional Forum," in *Imperfect Unions: Security Institutions over Time and Space,* ed. Helga Haftendorn, Robert O. Keohane, and Celeste A. Wallander (London: Oxford University Press, 1999), 287–324; and Michael Leifer, *The ASEAN Regional Forum: Extending ASEAN's Model of Regional Security,* Adelphi Paper No. 302 (Oxford: Oxford University Press, 1996), 22.

46. One example of China's foot dragging was its desire to change the wording of the evolutionary goals of the ARF, as stated in the 1995 Concept Paper, from "development of conflict resolution mechanisms" to "elaboration of approaches to conflict." For a more complete discussion, see Leifer, *The ASEAN Regional Forum: Extending ASEAN's Model of Regional Security*.

47. For a discussion of these joint declarations, see Jurgen Haacke, "China and ASEAN: Setting Parameters for Future Co-operation," in *Contemporary China: The Dynamics of Change at the Start of the New Millennium*, ed. P.W. Preston and Jurgen Haacke (London: Routledge, 2003), 264–269.

48. For the full text, see "Joint Declaration of the Heads of State/Government of the Association of Southeast Asian Nations and the People's Republic of China on Strategic Partnership for Peace and Prosperity" (October 8, 2003), available at www.aseansec.org/15266.htm.

49. According one Chinese scholar, signing the TAC was "not only a symbolic expression of a positive attitude, but a substantive political promise." See Chen Fengjun, "Jiaqiang Zhongguo yu Dongmeng Hezuo de Zhanlue Yiyi," 25.

50. China did not criticize Myanmar when a military junta took over control of the country in 1988 and subsequently jailed democracy activists, and it did not criticize Cambodia during its coup in 1997. Although China certainly had its own reasons for supporting autocrats, China's behavior also showed its respect for sovereignty and noninterference, principles which the ASEAN countries hold dear.

51. See M. Taylor Fravel, "Regime Insecurity and International Cooperation: Explaining China's Compromises in Territorial Disputes," *International Security*, vol. 30, no. 2 (Fall 2005): 46–83; and Henry Kenny, *Shadow of the Dragon: Vietnam, China, and the Implications for U.S. Foreign Policy* (Dulles, VA: Potomac Books, 2002).

52. See Ian James Storey, "Creeping Assertiveness: China, the Philippines and the South China Sea," *Contemporary Southeast Asia*, vol. 21, no. 1 (April 1999): 95–118. In April 2001, the Philippines also claimed that China had upgraded the communications facilities on Mischief Reef. See Carlyle A. Thayer, "Making the Rounds," *Comparative Connections*, CSIS, China-Southeast Asia Relations (2nd quarter, 2001).

53. According to the article, the parties pledged to "exercise self-restraint in the conduct of activities that would complicate or escalate disputes and affect peace and stability including, among others, refraining from action of inhabiting on the presently uninhabited islands." For the complete text, see "Declaration on the Conduct of Parties in the South China Sea" (November 4, 2002), available at http://www.aseansec.org/13163.htm.

54. For an optimistic assessment, see Mely Caballero-Anthony, "Major Milestone in ASEAN-China Relations," *IDSS Commentaries* (November 2002). For a more pessimistic assessment, see Ralf Emmers, "ASEAN, China and the South China Sea: An Opportunity Missed," *IDSS Commentaries* (November 2002).

55. See "Philippines, China, Vietnam to Conduct Joint Marine Seismic Research in South China Sea," Xinhua (March 14, 2005); and Luz Baguioro, "Three Nations Sign Pact for Joint Spratlys Survey," *Straits Times* (March 15, 2005).

56. See "Spratlys Complex," *Manila Standard* (August 8, 2005).

57. See Ang Cheng Guan, "Vietnam-China Relations Since the End of the Cold War," *IDSS Working Paper* (November 1998); and Kenny, *Shadow of the Dragon*.

58. For example, two Chinese Jianghu-class frigates and an intelligence ship were spotted off Scarborough Shoal in May 2001, but there was never any attempt to use

force. See Thayer, "Making the Rounds," *Comparative Connections,* CSIS, China-Southeast Asia Relations (2nd quarter, 2001).

59. See Luz Bagiuro, "Manila Considers Spratlys Protest," *Straits Times* (November 8, 2003).

60. See Leszek Buszynski, "Realism, Institutionalism, and Philippine Security," *Asian Survey,* vol. 42, no. 3 (May–June 2002): 483–501; and Ang Cheng Guan,"Vietnam-China Relations Since the End of the Cold War."

61. See "Matrix of ARF Decisions and Status, 1994–2004," available at www.aseansec.org/arf.htm; and Sheldon Simon, "Managing Security Challenges in Southeast Asia," *NBR Analysis* (July 2002).

62. See S.D. Muni, *China's Strategic Engagement with the New ASEAN: An Exploratory Study of China's Post-Cold War* (Singapore: IDSS, 2002). Out of sixteen outgoing "major military exchanges" from 2001–4, seven were with CLMV countries and four were with Thailand. Out of thirty-five incoming exchanges, eleven were with CLMV countries and ten were with Thailand. These figures come from appendices entitled "Major Military Exchanges with Other Countries" in China's *2002 and 2004 Defense White Papers*. The white papers are available at www.china.org.cn/e-white/.

63. See "China, Vietnam Hold First Round of Defensive Security Consultation," Xinhua (April 11, 2005); Luige A. del Puerto, "RP, China to Boost Defense Ties," *Philippine Daily Inquirer* (May 23, 2005); "China Offers Package Deal on Anti-Aircraft Missiles," *New Straits Times* (July 21, 2004); and Primastuti Handayani, "RI, China Seal Economic and Defense Deals," *Jakarta Post* (July 29, 2005).

64. The full texts of the white papers are available at http://www.china.org.cn/e-white/.

65. See "Iron Fist-2004 Expands China's Military Cooperation with Foreign Armed Forces," Xinhua (September 25, 2004).

66. For more on the efforts through the ARF, see Brad Glosserman, "The ARF Breaks New Ground," CSIS, *PacNet Newsletter*, no. 32 (August 9, 2002). For the text of the joint declaration, see "Joint Declaration of ASEAN and China on Cooperation in the Field of Non-Traditional Security Issues" (November 4, 2002), available at www.aseansec.org/13186.

67. See Michael Vatikiotis, "Military Alliances: A Diplomatic Offensive," *Far Eastern Economic Review* (August 5, 2004); and Lee Kim Chew, "China Could Play Part in ASEAN's Maritime Security," *Straits Times* (June 24, 2004).

68. See Joshua Kurlantzick, "China's Chance," *Prospect,* no. 108 (March 2005); Jehangir Pocha, "The Rising 'Soft Power' of India and China," *New Perspectives Quarterly,* vol. 20, no. 1 (Winter 2003); and Jane Perlez, "Across Asia, Beijing's Star Is in Ascendance," *New York Times* (August 28, 2004). The concept of soft power, as distinct from hard power, refers to getting what you want through attraction rather than coercion, especially the attractiveness of a country's culture or ideas.

69. See Eric Teo Chu Cheow, "China's Rising Soft Power in Southeast Asia," *CSIS Pacific Forum, PacNet*, no. 19A (May 3, 2004).

70. Given that many of the singers, actors, and directors are from Taiwan or Hong Kong, whether or not the popularity of these movies and singers should count as soft power for mainland China is still a debatable point.

71. See "Slowly but Surely, Mandarin Becomes Second Foreign Language," *Jakarta Post* (March 12, 2005); and Teo Cheng Wee, "Hooked on Chinese," *Straits Times* (November 28, 2004).

72. For examples of this view, see Dana R. Dillon and John J. Tkacik, Jr., "China

and ASEAN: Endangered American Primacy in Southeast Asia," *Heritage Foundation Backgrounder*, no. 1886 (October 19, 2005); James Steinberg's statements in Michael Vatikiotis and Murray Hiebert, "How China Is Building an Empire," *Far Eastern Economic Review* (November 20, 2003); Marvin Ott's statements in Stuart Grudgings, "China Shows Other Asian Nations It Can Give as Well as Take," *Financial Express* (October 9, 2004); and James A. Kelly, Assistant Secretary of State for East Asian and Pacific Affairs, "An Overview of U.S.-East Asia Policy" (Testimony before the House International Relations Committee, June 2, 2004).

73. A recent analysis of China's influence argues that China still lacks conduct-shaping and context-shaping power in Southeast Asia. See Jurgen Haacke, "Seeking Influence: China's Diplomacy Toward ASEAN After the Asian Crisis," *Asian Perspective*, vol. 26, no. 4 (2002): 13–52.

74. For Japanese figures, see www.asean.or.jp/general/statistics. For U.S. figures, see http://www.us-asean.org/statistics/.

75. From 1990 to 2001, cumulative U.S. FDI in ASEAN reached $41.1 billion and cumulative Japanese FDI reached $63.7 billion. See ASEAN, "Statistics of Foreign Direct Investment in ASEAN: Comprehensive Data Set," 2002 edition, available at www.aseansec.org/14549.htm. In terms of total FDI to ASEAN from 1995 to 2001, Japan accounted for 21.6 percent, the United States accounted for 14.5 percent, Taiwan accounted for 6.3 percent, and China only accounted for 0.9 percent.

76. See "U.S. Investment in ASEAN," available at www.us-asean.org/statistics/US_investment.htm.

77. For discussions of Japan's assistance during the financial crisis and trade, investment, and ODA relations with ASEAN, see Sueo Sudo, *The International Relations of Japan and Southeast Asia: Forging a New Regionalism* (London: Routledge, 2002). For the specific claim of $23 billion in ODA, see Brad Glosserman, "Devil of Amity Lurks in Free Trade Details," *Japan Times* (December 21, 2003).

78. See Roger Mitton, "Debate Rages on over Lax U.S. Policy on ASEAN," *Straits Times* (February 10, 2006).

79. For an excellent analysis of these recent developments in U.S.-ASEAN military relations, see Sheldon Simon, "Southeast Asia: Back to the Future," in *Strategic Asia 2004–05: Confronting Terrorism in the Pursuit of Power*, ed. Ashley J. Tellis and Michael Wills (Seattle: NBR, 2004).

80. See Secretary Colin L. Powell, "Briefing on Trip to East Asia" (July 29, 2001), available at http://www.state.gov/secretary/rm/2001/4347.htm.

81. This strategy of maintaining a balance or "counter-dominance" strategy, sometimes also called hedging, is widely recognized by analysts of Southeast Asia. See Acharya, "Seeking Security in the Dragon's Shadow": and Evelyn Goh, "Meeting the China Challenge: The U.S. in Southeast Asian Security Strategies," *Policy Studies 16*, East-West Center, Washington, D.C. (2005). Chinese analysts also recognize this strategy. For example, see Zhang Xizhen, "Dongmeng de Daguo Junshi Zhanlue" (ASEAN's strategy of maintaining a balance among the great powers), *Guoji Zhengzhi Yanjiu* (Studies of international politics), no. 2 (1999): 120–127.

82. For the complete statements, see "Chairman's Statement of the Ninth ASEAN Plus Three Summit," Kuala Lumpur (December 12, 2005), available at www.aseansec.org/18042.htm; and "Kuala Lumpur Declaration on the East Asia Summit," Kuala Lumpur (December 14, 2005), available at www.aseansec.org/18098.htm.

83. See "China Supports ASEAN's Leading Role in East Asian Cooperation: Chinese FM," Xinhua (December 10, 2005).

84. For a short discussion of Zoellick's visit, see Evelyn Goh, "Renewed American Diplomacy: Keeping Southeast Asia on the U.S. Radar Screen, *IDSS Commentaries* (May 24, 2005).

85. See Harvey Kaur, "Rice's Absence at ARF Sends Wrong Signals," *New Straits Times* (July 29, 2005).

86. In the first three months after the early harvest program began, China's exports to Thailand increased 200 percent, while Thailand's exports to China only increased 80 percent. See Michael Vatikiotis, "A Too Friendly Embrace," *Far Eastern Economic Review* (June 17, 2004); and "Benefits of China FTA Downplayed," *Nation* (January 26, 2005). Due to fears of being flooded by Chinese agricultural products, the Philippines pulled out of the early harvest program. See Elaine Ruzul S. Ramos, "RP Nixes ASEAN-China Trade Deal," *Manila Standard* (April 18, 2005).

87. See Walden Bello, "China-ASEAN Free Trade Agreement: ASEAN is Sadly Lacking in Relevance," *Bangkok Post* (December 14, 2004).

88. The fish catch in Cambodia dropped by half in 2003 and the fish catch in northern Thailand declined by half from 2000 to 2004. See Peter S. Goodman, "Manipulating the Mekong," *Washington Post* (December 30, 2004); Jane Perlez, "In Life on the Mekong, China's Dams Dominate," *New York Times* (March 19, 2005); and Evelyn Goh, "China in the Mekong River Basin: The Regional Security Implications of Resource Development on the Lancang Jiang," *IDSS Working Paper* (July 2004). In addition, China's ability to control the amount of water that goes downstream gives it a level of political control that may also be troubling to countries worried about the possibility of China trying to treat them like colonies in the future.

89. For more details, see "'Cruel Oil' Signals Doom for Borneo's Eden," *South China Morning Post* (February 6, 2006).

90. See Ronald Montaperto, "Smoothing the Wrinkles," *Comparative Connections,* CSIS, China-Southeast Asia Relations (2nd quarter, 2004).

91. See "Mei Yue Hezuo Yiweizhe Shenme" (What does U.S.-Vietnamese cooperation mean?), *Huanqiu Shibao* (Global times) (June 27, 2005); author's interviews, Beijing (summer 2005).

92. For discussions of U.S. motives in the war on terror, see Yao Jianguo, "Mei Jie Fankong Jiakuai zai Dongnanya Junshi Chongfan" (America's taking advantage of the opportunity of opposing terrorism to accelerate its military return to Southeast Asia), *Shijie Zongheng* (The contemporary world), no. 8 (August 2002): 19–20; and Xia Liping, "Meiguo 'Chongfan Dongnanya' jiqi dui Yatai Anquan de Yingxiang" (America's 'return to Southeast Asia' and its influence on Asia-Pacific security), vol. 8 (August 2002): 18–22. For discussions of motives in response to the tsunami, see "Haixiao Jiuyuan gei Shijie yi ge Jihui" (Tsunami assistance gives the world an opportunity), *Huanqiu Shibao* (Global times) (January 26, 2005); and Xiao Ding, "Politics Surrounding the Tsunami," *Beijing Review* (February 24, 2005).

93. See Yin Chengde, "Dongmeng Waijiao Xin Zouxiang: Jian lun Zhongguo yu Dongmeng Guanxi" (New trends in ASEAN's diplomacy: Sino-ASEAN relations), *Guoji Wenti Yanjiu* (International studies), no. 3 (2004): 24.

III

Conclusions

8

Evaluating China's Strategy Toward the Developing World

Eric Heginbotham

Chinese diplomacy appears to be taking the developing world by storm. Its leaders seem to be everywhere: signing investment agreements, building roads, forming "strategic partnerships," and gaining membership in new or expanded regional organizations. It has burnished its image by dispatching blue-helmeted Chinese soldiers and policemen on United Nations (UN) peacekeeping missions, donating money and equipment for disaster relief efforts, settling most of its border disputes, and engaging actively in a host of multilateral organizations around the world. It has also overhauled its foreign policymaking machinery, enabling political, bureaucratic, business, and academic experts to function as part of a more seamless whole.

The transformation of China's foreign policy from one that was largely ideologically driven—and often hapless—during the 1960s and 1970s, to the proactive and adept approach witnessed over the last several years, took place in two phases. The first, rapid and revolutionary, occurred with the ascendance of Deng Xiaoping after 1978 and his introduction of a pragmatic foreign policy, within which economic ends and means took pride of place. The second dates to the mid- or late 1990s, and came largely in response to new challenges: growing concerns in foreign capitals over the direction of Chinese power, Beijing's concerns about U.S. primacy, and the increasing demands of supporting a large and increasingly sophisticated economy. China's current approach emphasizes regional and global multilateralism, confidence-building with states on its periphery, the promotion (where possible) of a multipolarity, and the deepening of comprehensive ties with key partners around the world.[1]

For the most part, Chinese foreign policy has proven adaptive, enabling China to march forward while avoiding the pitfalls inherent in the rise of major powers. Tensions with most of its neighbors in East Asia, South Asia, and Central Asia have eased considerably, and a number of those states have become political—as well as economic—partners. Beijing's diplomacy has facilitated the development of new trade outlets in Latin America, Africa, and the Middle East. China's purchase of equity stakes in upstream energy and mineral projects, as well as the transportation infrastructure to facilitate extraction, may help it secure the raw materials necessary to power its expanding industrial production.

But while China's new diplomacy is more adept than anything seen from Beijing in past decades, the nature and extent of its influence is still limited. Its new approach has mitigated many of the immediate concerns of its neighbors about its political and military direction, but questions about its behavior in the long term linger. Its economic diplomacy is astute, but with trade growing by leaps and bounds (an average of 15 percent annually in real terms between 1999 and 2005), the demands on this diplomacy continue to grow. At a more fundamental level, tensions also exist in Beijing's objectives. Its desire to be recognized as a constructive member of international society bumps up against resource diplomacy that supports autocratic and abusive governments in several corners of the world. Beijing also occasionally makes rudimentary diplomatic mistakes: leading up to the inaugural East Asian Summit, it pushed for an exclusive approach to membership that had little support from most regional states. As the world changes—and China's place within it—Beijing's foreign policy must continue to evolve or face the danger of becoming dysfunctional.

Questions regarding Chinese objectives and their compatibility with the existing world system abound. Will Beijing employ its new power and influence to advance its position at the expense of others, or will it work toward win-win solutions to international problems? Will China undermine international law and challenge evolving global norms and institutions, or will it redefine its interests to include support for these norms? In what ways will it try to shape new norms, and how successful will it be in those efforts?

We cannot know the answer to these questions about China's future trajectory, but we can look at recent developments, and those developments suggest a mixed record. Chinese leaders have, in some cases, proven capable of reevaluating their definition of national interests, particularly when their preexisting trajectory seems to be leading them toward conflict or isolation. They are now strongly committed to international stability, without which they believe that China's ambitious economic goals cannot be met. Their ideas on arms control, multilateralism, and many areas of international law have also edged toward "Western" norms. On other issues, Chinese views have been significantly more resistant to change. Beijing's attitude toward human rights and the sanctity of sovereignty has evolved only slowly, widening the gap between Beijing and most other major states, where norms on these questions have changed more quickly.

This rest of this chapter is divided into four sections. The first briefly outlines the evolution of Chinese policy toward the developing world during the era of Deng Xiaoping, from an ideologically oriented foreign policy to one oriented around economic interests. The second describes the tensions that were built into Beijing's foreign policy during the 1990s and resulted in the shift toward a more proactive and politically engaged foreign policy after 1997. The third summarizes the characteristics or elements of China's new diplomacy as they relate to Beijing's activities in the developing world. And the fourth offers tentative answers to several key questions: How does China define its national interests in these areas and how has that definition changed over time? How effectively do its policies achieve or support those interests? What is the impact of Beijing's foreign policy on global and regional stability and emerging international norms? And what are the prospects for future changes in Chinese thinking about and policy toward the developing world?

From Ideology to Economics, 1978–1997

Until the late 1970s and Deng Xiaoping's "reform and opening," Beijing's policies in the developing world were ostensibly driven by ideological concerns: unity with the "Third World" and support for "national liberation."[2] A strict dichotomy between ideology and interests would be misleading. Beijing often used ideology as a means to pursue its competition with other states—the United States during the 1950s and the Soviet Union during the 1960s and 1970s.[3] Whether ideological or interest-driven, China's foreign policy prior to 1978 was oriented around political objectives (for more on this period, as well as China's approach to foreign policy as far back as Imperial times, see the chapter by Derek Mitchell and Carola McGiffert in this volume).

The late 1970s and early 1980s saw China set on a new domestic and international course. With the rise of Deng Xiaoping, China dedicated itself to the "four modernizations," and economic interests gained primacy in Beijing's foreign policy. Despite continuing tensions over Cambodia and Afghanistan, Beijing downgraded the threat posed by the Soviet Union and the probability of world war. Economically, China's most urgent need was for foreign technology from the developed world. Financing that technology required hard currency, the cultivation of new export markets, and budgetary discipline in other areas. In the developing world, cultivating ties with important economic suppliers and markets replaced Beijing's former emphasis on underwriting revolution or countering Soviet initiatives.

Beijing did not abandon ideological appeals or balance of power politics entirely. In 1982, it declared an "independent foreign policy" with which it distanced itself from the United States, at least rhetorically, and claimed ideological leadership of the Third World. But in practice, it took a pragmatic attitude toward cooperation with the United States and the Soviet Union. In the developing world, its new focus on economically viable and important relationships provided a more realistic foundation for the future growth of Chinese influence. The shift was reflected in Beijing's foreign policy lexicon. By the late 1980s, the use of the term "developing states," an economic descriptor, had largely replaced the moniker "Third World," a term with an explicitly political definition.[4] The rebuilding of academic institutions decimated during the Cultural Revolution, and the creation of new organizations (including the Chinese Academy of Social Sciences in 1977), helped begin to fill the knowledge vacuum about the outside world.[5]

China's foreign policy remained hampered throughout the 1980s by a number of limitations.[6] Despite trumpeting the merits of south-south cooperation, many states had little to offer in the way of mutually profitable trade or technical cooperation. This was especially true among those states with which the China had enjoyed close ties during the 1960s and 1970s, including several states in sub-Saharan Africa.[7] Needing to import expensive capital equipment, and therefore facing tight budget constraints, Beijing was handicapped by its inability to continue supplying foreign aid. To the extent aid was offered, it was now expected to provide mutual economic benefits.[8] China did enjoy a boom in arms sales during the 1980s, especially to the Middle East,[9] but those sales came at the cost of increased international suspicion about China's motives and direction.

Regardless of its handicaps, however, Beijing's foreign policy scored important successes during the decade. In its relations with developing states, gains within Asia were particularly prominent. Prompted by the Vietnamese invasion of Cambodia in 1978, Beijing cut off support for communist guerrillas in Thailand and moved to establish a strategic relationship with Bangkok. The more general shift away from ideological foreign policy and the expansion of economic ties underpinned improved ties with Malaysia, the Philippines, and Indonesia. Overall, two-way trade with the developing states of Asia grew some 440 percent (in constant dollar terms) between 1978 and 1991, significantly faster than China's total global trade (at 250 percent).[10] Another notable bright spot included Latin America, where the diminution of Beijing's revolutionary rhetoric was particularly welcome in the region's large and generally conservative regimes.[11]

Chinese foreign policy during the early and mid-1990s followed the same pragmatic, if not necessarily proactive, logic it had followed during the 1980s. China's isolation immediately following the Tiananmen massacre sparked a brief look at greater reliance on relationships in the developing world. Beijing's options, however, remained limited by its need for foreign direct investment (FDI) from the West, and its leaders correctly concluded that sanctions would not remain in place. In many respects, China's pragmatic approach remained successful. During the 1990s, it normalized relations with Singapore, Indonesia, Saudi Arabia, Vietnam, Israel, and South Korea, among others. Trade ties in virtually every region were extended and deepened.

New Challenges and Opportunities in the Mid-1990s

Despite its successes (and in some cases because of them), China was confronted by a new set of challenges and opportunities by the mid-1990s that appeared to demand more than simple economic pragmatism. As the international environment changed, Beijing's existing foreign policy looked increasingly inadequate.

As Michael Glosny observes in chapter 7, nowhere was the inadequacy of Beijing's existing foreign policy more evident than in China's relations with its Southeast Asian neighbors. China's growing—although still limited—military power, combined with military encroachment in the South China Sea's Spratly Islands and its refusal to undertake multilateral negotiations on the issue, prompted unease throughout the region. Crises across the Taiwan Strait and Chinese missile tests there helped drive concerns to new heights in 1995–96. Normalization with individual states and Beijing's half-hearted attempts to engage ASEAN without addressing its security concerns were inadequate to counter the slide in perceptions.[12] Several ASEAN states began to hedge their bets. The Philippines moved to reenergize its treaty with the United States, and Vietnam invited U.S. military officials to tour military facilities in 1997.[13] ASEAN moved to admit Vietnam, Laos, Myanmar, and Cambodia on an accelerated schedule.[14] In a remarkable departure from its past practices, Indonesia signed a defense agreement with Australia in 1996.

Growing regional suspicions were particularly troublesome in the context of rising tensions with the United States. The United States-led NATO operations in the Balkans led Beijing to question whether America might indeed use its position as the world's sole remaining superpower to make itself the arbiter of international justice.[15] The 1999 bombing of the Chinese embassy in Belgrade was a watershed. Chinese media began discussing American *baquan zhuyi* ("hegemonism") with an intensity not seen since the 1970s.[16] Even during the worst of times, Beijing sought to maintain cordial relations with the United States, acknowledging publicly that stable Sino-American relations are critical to achieving the nation's larger economic and political objectives. American primacy, and the specter of its use against China, has nevertheless led Beijing both to continue working to improve the bilateral relationship and to hedge its bets wherever possible.

China's growing economy and trade gave it greater international weight by the late 1990s, but it also resulted in growing economic frictions. China's imports from the developing world consistently surpass its exports to it.[17] But many of China's trade partners in the developing world expressed concerns about the composition of trade and the impact on domestic industry. China's primary imports from many of these states are raw materials, while its exports are largely manufactured goods. Concerns about the impact on local economies—and industry in particular—are particularly prominent in Latin America. By the late 1990s, Latin American anti-dumping measures against China—some of the toughest in the world—represented a serious obstacle to the further development of trade.[18] Partners in the Middle East and Southeast Asia, though, shared similar concerns.

Economic growth has also driven a voracious appetite for raw materials, driving a new need for intensified management of diplomatic relations with key suppliers. China is currently the world's leading consumer of cement (or around 50 percent of the total), coal (30 percent), steel (36 percent), and copper (17 percent).[19] In 2004, it surpassed Japan as the world's second largest consumer of oil. An increasing share of its raw materials is imported. China, a net oil exporter before 1993, now imports more than 40 percent of its oil from abroad.[20] China's growing demand for raw materials, combined with rising demand from India, has driven up the global price for many of these commodities. In 2004, robust Chinese demand for iron ore and coking coal enabled producers in Australia to secure contract price increases of 71 and 120 percent, respectively.

Although hardly a new challenge, Beijing also continued to face competition for international recognition with Taiwan. A balanced assessment of net gains and losses would almost certainly credit China as the winner during the 1990s. During the early part of the decade, Taiwan lost the recognition of Saudi Arabia and South Africa—the last two significant powers in Taipei's camp. Nevertheless, from a strictly numerical standpoint, Taiwan improved its position between 1980 and 1995, providing added incentives for Beijing to increase its presence in Africa, Central America, and the Caribbean (where the bulk of states that recognize Taiwan are located).

Finally, new opportunities were also important in influencing China's redefinition of its national interests and in reshaping China's thinking about how it might meet its new challenges. In late 1997, ASEAN sought the partnership of China, Japan, and South Korea to deal with the fallout from the financial crisis that was then underway. In addition to discovering ASEAN's openness to a leadership role for China, it also found that relatively modest financial contributions or sacrifices, under the right circumstances, could yield substantial goodwill—as its $300 million donation to Indonesia and its steadfast refusal to devalue its currency did during the 1997–98 crisis. Beijing's early success with the Shanghai Five grouping (which later became the Shanghai Cooperation Organization) also suggested the potential benefits of multilateralism.

China's New Diplomacy

In response to these challenges and opportunities, China has adjusted its foreign policy since the mid- to late 1990s. Its new diplomacy is more proactive and flexible. Beijing has embraced multilateralism and confidence-building measures to mitigate security concerns. It has identified a number of states and regional organizations as "strategic partners." It has sought to increase its equity stakes in resource exploitation. Although less prominent in China's foreign policy than during the 1970s, economic assistance is employed strategically to further bilateral relationships. Beijing now emphasizes and cultivates soft power. For the most part, there is synergy between these various aspects of China's new diplomacy, but there are also some elements that create tensions with one another.

Although the changes in Chinese foreign policy after 1997 are generally evident in China's foreign policy, they are most apparent—and most momentous—in Beijing's approach to the developing world. This is where the raw materials to fire China's economy can be found in greatest abundance. It is where new markets can be found for Chinese products—markets where brand-name loyalties are not as firmly rooted as they are in the developed world. It is also where Beijing has the greatest room to maneuver politically, among a variety of states that share its desire to see a more multipolar world, one where American power is constrained collectively. The developing world is hardly a single entity, and part of China's new thinking is a differentiated approach to different types of developing states. The elements of China's new diplomacy are discussed below both in general terms and in their specific application to the developing world.

New Thinking and Concepts

New thinking on China's relations with the outside world has been reflected in statements by officials and scholars. There are, of course, important elements of continuity. Beijing's diplomacy continues to sing the praises of the "five principles of peaceful coexistence," as it has since 1954 when Zhou Enlai first used the term during a visit to India and Burma.[21] Despite elements of continuity, though, new formulations and debates since the late 1990s suggest shifting thinking on foreign policy, or at least expanded space for rethinking foreign policy.

In 1997, Foreign Minister Qian Qichen and others began to outline a "new security concept." The concept contains a number of propositions about the causes of peace and international stability that are, broadly speaking, consistent with liberal views of international relations.[22] To be sure, national sovereignty, embodied in the five principles of peaceful coexistence, would not be compromised. But traditional power politics are rejected as counterproductive and destabilizing. True and lasting security is seen as mutual. In dealing with conflicts between states, discussion and cooperation advance mutual understanding and underpin stability. The merits of multilateralism are highlighted. Several of these themes are also consistent with Beijing's discussion of "China's peaceful rise," which proposes that China's rise (1) depends on its own continued economic reform, (2) will take a long time, and (3) will not and could not be accomplished at the expense of other nations.[23]

Like all other Chinese foreign policy formulations (including "peace and development" or "democracy in international relations"), the new descriptive phrases and concepts do not represent fully articulated theories or policies. Rather, they signal a leadership emphasis on particular priorities or at least advocacy of those priorities by an important subsection of the leadership.

Multilateralism and Confidence Building

Multilateralism has become a keystone in Beijing's new approach to international relations, particularly in its relations at the regional level with developing states. Its application began haltingly and initially with states on China's periphery (*zhoubian guojia*). In 1995, China initiated a Shanghai Five dialogue—including Russia, Kazakhstan, Kyrgyzstan, and Tajikistan, in addition to China—to discuss confidence-building measures along their respective borders. As Matthew Oresman suggests in chapter 3, the scope of the group's discussions expanded quickly. By 2000, the five countries had concluded a series of agreements on troop reduction in border areas, the mutual observation of military exercises, border demarcation, and cooperative efforts to combat cross-border crime and terrorism. The success of the group, which added Uzbekistan, and became the Shanghai Cooperation Organization in 2001, made it a model for China's subsequent multilateral efforts and is often used to illustrate principles in the nation's new security thinking.[24]

In Southeast Asia, China's engagement with ASEAN now underpins its relations with member states. In 1994, three years after establishing relations with ASEAN, China joined the new Asian Regional Forum (ARF), a body that discussed security issues but was hardly action oriented. The ASEAN+3 grouping (which included ASEAN, China, Japan, and South Korea) was launched at ASEAN's behest in 1997. In late 2000, China began discussing a separate "bilateral" free trade agreement (FTA) with ASEAN. ASEAN, however, remained concerned about Beijing's intentions in the South China Sea. Negotiations on the two issues proceeded together, and in November 2002, China and ASEAN signed both a declaration of conduct in the South China Sea and the China-ASEAN Free Trade Agreement.[25] In both Central Asia and Southeast Asia, China's interest in improving the tenor of its international relationships has trumped its desire to pursue long-held territorial claims.[26]

China's engagement of regional groupings may have begun in Asia, but it now extends well beyond its community of "peripheral states." In Latin America, it gained permanent observer status in the Organization of American States (OAS) in May 2004, and it has applied for the same status in the Inter-American Development Bank. Beijing and the Caribbean states held the first meeting of the China-Caribbean Economic Forum in February 2005.[27] In the Middle East, Beijing and the 22-member Arab League established the Sino-Arab Cooperation Forum in January 2004, committing all sides to ministerial-level meetings. It is negotiating an FTA with the six states of the Gulf Cooperation Council. The Forum on China-Africa Cooperation (FOCAC), established in 2000, facilitates both Chinese diplomacy and economic activity on the continent of Africa. In November 2006, China hosted FOCAC's first summit meeting (and third ministerial meeting), an event that saw heads of state or government leaders from more than forty African states gather in Beijing.

A Differentiated Approach to the Developing World

The advanced industrial world still receives top billing in China's hierarchy of importance. In Jiang Zemin's report to the sixteenth Party Congress in November 2002, precedence is assigned in the following order: developed states, neighboring countries, and other developing states.[28] The developed world is still the major source of China's incoming FDI and imported technology, and as of 2005, trade with the developed world still accounted for more than half of China's total trade volume. Nevertheless, the relative importance accorded the developing world is unmistakably rising. Relations with the developing world have once again become a prominent topic in the public discourse on Chinese diplomacy.[29]

In part, the renewed emphasis reflects new economic realities. After remaining relatively constant between 1978 and 1991, the percentage of China's trade conducted with the developing world rose from 34 percent of its global total in 1991 to 47 percent in 2005. The new emphasis also reflects new political calculations. Romantic notions about the inevitable rise of "the South" died long ago, but Beijing now recognizes a differentiated developing world. Some areas of the developing world are not merely lagging, but positively dropping into irrelevance. A few "large [developing] states or groups of states," however, are growing faster than even the advanced industrial states and "will occupy important positions in the future multipolar order."[30] These states are potential political and economic partners.

Strategic Partnerships

China has developed declared partnerships of various types with a number of states—some developing, some developed. In some cases, these partners are political heavyweights. In others, they are critical to China as sources of scarce raw materials. In the Chinese diplomatic lexicon, the most significant of these are "strategic partnerships" (*zhanlue huoban guanxi*).[31]

In the developing world, strategic partnerships have been signed with Brazil (1993), Venezuela (2001), Mexico (2003), South Africa (2004), Argentina (2004), India (2005), Kazakhstan (2005), Indonesia (2005), Nigeria (2006), and Algeria (2006).[32] China has also signed strategic partnerships with regional groupings of developing states, including, in the developing world, Africa (2000) and ASEAN (2003). China, of course, maintains long-standing security relationships with Pakistan and North Korea, though relations with these two countries are complicated by Beijing's larger economic (and arguably political) interests in Pakistan's rival, India, and North Korea's opposite, South Korea. (In the case of Pakistan, the relationship is now described as a long-standing strategic partnership, perhaps to mitigate sensitivity to China's new strategic partnership with India.) China also enjoys close strategic relations with some smaller states that may not rise to the implied peer status suggested by a formal partnership, including Burma, Sudan, Cuba, and Ghana, among others.

Anchoring foreign policy in the developing world to a handful of partners allows China to focus its limited diplomatic resources on developing comprehensive relationships with the states of most potential value. Specific motivations appear to vary. Venezuela and Russia share interests in "promoting multipolarity" and "democracy in international relations"—code phrases for collective soft balancing against U.S. hegemony. More commonly, China's political ambitions in these strategic partnerships are limited to more defensive ends, ones that might be generalized as counter-containment.[33] Whether or not Washington's strategy vis-à-vis Beijing includes elements of containment—an assertion strongly and consistently denied by the U.S. government—uncertainties about both present and future U.S. intentions shade the calculations of Chinese policymakers. As Rollie Lal argues in chapter 6, the Chinese partnership with India can best be understood as an attempt to prevent India from being wooed successfully by the United

States as a collaborator against Beijing. The same appears true in the case of China's improved relations with the Southeast Asian states. China's strategic partnerships are not alliances in the traditional sense of the word, and therefore do not require partner nations to take sides between the United States and China—something most of China's strategic partners would be unwilling to do in any case.

Evaluating the relative weights of political and economic motivations in Beijing's partnerships is difficult, and clearly the two tend to work in tandem. Nevertheless, it is probably fair to say that economic interest provides a more common theme. Apart from the pursuit of specific trading interests, depending on the economic assets of the country in question, Beijing has, in all its negotiations over strategic partnership, pushed for the formal recognition of China's "market economy status"—a status that makes it more problematic, under WTO rules, to impose anti-dumping penalties.

Investment and Equity Stakes

China's outward investments, particularly in the energy field, have been among the most dramatic aspects of its new diplomacy. The image of Hu Jintao, with seemingly bottomless pockets, marching into foreign capitals and committing Chinese investment to the tune of billions of dollars has generated admiration and concern in different parts of the world. Despite the impression of power and authority, however, the efficacy of China's hunt for equity stakes is uncertain.

China's FDI appears to be designed to mitigate concerns over the impact of trade on China's partners, to hedge against possible restrictions and tariffs on Chinese imports, and, most importantly, to address China's own resource shortages. In the first role, investment by quasi-state companies and backed by state banks is a convenient policy tool. China's outgoing FDI remains a fraction of its incoming FDI (according to Chinese statistics: $12.3 billion vs. $72.4 billion in 2005).[34] But outgoing FDI grew far faster than incoming FDI during 2003–2005 (340 percent vs. 29 percent, respectively), and the political significance of even limited outward investment in poor or financially troubled countries can be enormous. When Hu Jintao visited Buenos Aires in November 2004, he brought with him an investment exploration package totaling almost $20 billion.[35] Not surprisingly, Argentina—struggling after its 2001 default—welcomed the promised investment and reciprocated by recognizing China's market economy status.

Although outward FDI in energy and minerals sometimes supports China's trade and political agenda, its primary purpose is to secure stable access to scarce raw materials. After moving toward greater reliance on international markets during the 1980s, China's extractive industries appear to have moved sharply back in the other direction—toward investments in equity stakes—over the last several years.[36] Countries hosting significant Chinese energy investment include Kazakhstan, Nigeria, Iran, Sudan, Angola, Russia, Ecuador, Venezuela, and Brazil, among others.[37]

China's outward investments are potentially powerful political and economic tools. There are, however, a number of questions about their actual efficacy. What will happen to these investments—and the relationships built upon them—in the event of a recession in China or a decline in global commodity prices? Backing by China's state banks may provide certain advantages for the oil companies, but it could also threaten the banking system itself if prices collapse. Spikes or sustained price increases could also pose challenges by providing incentives for more countries to follow Bolivia's lead and nationalize foreign-invested resource projects. And even in the absence of such developments, some of Beijing's recent investment agreements appear questionable. Many of the countries with which China has recently committed large investments are states in which it has limited experience and knowledge. Moreover, its own preparation to execute deals appears, in some cases, suspect. The primary contractor in the exploratory energy agreement with Argentina is Sonangol—a Chinese-Angolan holding company with a history of misplacing money.[38]

Trade

Although trade friction provided a key motivation for Beijing's more energetic diplomacy, deepening trade ties and improving the terms of trade also serve as one part of the solution. Despite concerns about the health of its own agricultural sector, Beijing has been willing to open this sector to competitors with comparative advantages in farm production. Early harvest agreements with ASEAN states made the 2002 China-ASEAN FTA more palatable.[39] Concessions on the opening of China's soybean market helped pave the path to Beijing's strategic partnership with Argentina in 2004.[40] In 2005, Beijing dropped tariffs on 190 commodities from twenty-five African nations. China's flexibility on trade issues, particularly on agricultural produce, has provided it with a major advantage over Japan in its negotiations for FTAs in Southeast Asia and elsewhere. Although China's splurge in outward investments has captured recent headlines, the steady, if less sexy, deepening of its trade ties probably provides a firmer basis for the development of its long-term political ties.

Economic Aid

In Beijing's bilateral diplomacy, economic aid does not play the central role it did during the 1960s and 1970s. Nevertheless, after two decades of minimal giving, aid is again recognized as an important diplomatic tool.[41] According to Chinese statistics, Beijing gave a total of $604 million in 2002 (the first year for which statistics on aid were published), $630 million in 2003, and $731 million in 2004.[42] Some outside observers provide higher numbers and suggest that aid budgets have grown steadily over the last decade.[43] In some instances, such as Beijing's aid to Indonesia and Thailand during the 1997–98 financial crisis and its relief effort following the December 2004 tsunami, aid appears intended to support the nation's general public diplomacy effort.[44] More often, however, it is directed at cementing ties with specific economic or strategic partners.

Arms Sales

Chinese arms transfers have declined substantially since the 1980s, but are still employed strategically. The total value of Chinese arms sales (in current dollars) fell from a high of around $5 billion in 1987 to an average of $600 million between 2000 and 2003—or from 11.5 percent of the global arms market to just 3.1 percent.[45] The merits of "simplicity" and "reliability," heralded as advantages of Chinese weapons during the Iran-Iraq War (1980-1988), lost their luster after American victories in the Gulf war and the Kosovo campaign. Some portion of China's arms sales, however, goes to poor, international pariah states, where the "enemies" are often lightly armed fighters or unarmed civilians. As Joshua Eisenman observes in chapter 2, the military and political value to these states is high.[46] China's white paper on Africa specifically highlights Beijing's commitment to "military-related technological exchange and cooperation" and the "army building of African countries for their own security." It remains to be seen whether Beijing might come to realize that its military diplomacy with pariah states cuts against its other interests, especially its efforts to cultivate its image and soft power with a broader range of states.

Soft Power

Perhaps the most surprising development in China's foreign policy has been the rise of its soft power and the government's new appreciation of it. China's soft power today—unlike during the Cold War—is largely generated spontaneously by its own society and semiautonomous institutions. Tourism has boosted China's visibility. In 2003, the number of Chinese tourists surpassed that of Japan, making them the second largest national group after Americans.[47] Chinese universities hosted 141,087 students in 2005, up from 77,628 in 2003, creating a future constituency for China overseas. Most foreign students now pay their own way, though the government sponsored 7,218 students in 2005—mostly from poor countries.[48]

Some efforts to generate soft power involve a more active state role. In chapter 5 Mao Yufeng describes Beijing's multifaceted public diplomacy efforts in the Middle East. There, Beijing cultivates an image as a longtime supporter of Arab causes. It has financed a number of exchange programs throughout the region, and established a Chinese Cultural Center in Cairo in 2002. In perhaps its most innovative move (especially in light of the secular nature of China's state), Beijing has capitalized on its 20 million Muslims to propagate the image of an Islamic friendly state, and it has facilitated exchanges between its own Muslim population and those of the Arab states. Chinese efforts in the Middle East are not unique. Similar patterns, adjusted for regional differences, are evident in all regions. China is establishing one hundred Confucius Institutes, mainly in East Asia, to promote the study of Chinese language and culture—and it had already constructed thirty such institutes by March 2006.[49]

International Organizations and Peacekeeping

Since 1978, Beijing has joined the world's organizing bodies and, to varying degrees, adjusted its behavior to conform to the responsibilities incurred.[50] It has not, however, placed a high premium on assuming financial or intellectual leadership in global bodies. For the most part, its engagement with *regional* organizations, like ASEAN+3 and the Asian Development Bank, has been more proactive than its involvement with *global* organizations. Some recent measures may, however, indicate increased attention to global organizations. In 2001, Beijing agreed to increase its annual quota to the IMF's Special Drawing Rights fund by 35 percent.[51] In 2003, the first UN agency ever headquartered on Chinese soil opened.[52] And in 2004, Beijing announced it would provide its first million dollars to the International Atomic Energy Agency's special fund for technical cooperation and enhanced nuclear security.[53] Given the small scale of these efforts, however, it would be premature to conclude that Beijing has made leadership in global organizations a high priority.

One important exception to China's general lack of enthusiasm for global activism is UN peacekeeping. In 1990, China dispatched its first peacekeepers to the Middle East. Between 1990 and mid-2006, it dispatched a total of 5,872 personnel on 15 UN peacekeeping missions, making it the largest contributor to peacekeeping operations among the permanent members of the UN Security Council.[54] As of November 2006, 1,648 Chinese peacekeepers were serving on ten missions.[55] Although some see sinister motives in Chinese peacekeeping, Beijing's new activism on the peacekeeping front came in response to years of encouragement by the United Nations, as well as by the U.S. government.[56] China has engaged in a broad range of operations, from Bosnia to Liberia to Cambodia. In most cases, whatever the political advantages it may derive, like those gained by other contributors, they are largely intangible and indirect.

Assessing China's Policy Toward the Developing World

How are we to assess the ultimate meaning and implications of these disparate elements of Chinese foreign policy in the developing world? Any summary evaluation should address the four questions laid out in the introduction to this book. First, what do China's declared policies and its actual behavior suggest about how it defines its national interests and how that definition has changed over time? Second, how effectively have its policies addressed those objectives? Third, what is the impact of Beijing's foreign policy on the world, particularly on international stability, on the U.S. position in the region and the world, and on international norms of corporate and government behavior? And fourth, what are the prospects for future change, both in Beijing's definition of its ends and means, and how might the United States influence its evolution?

China's Evolving Definition of National Interests

China's core objectives appear largely unchanged, but it has spelled out new intermediate objectives and sharpened its thinking on priorities in ways that have significantly altered the balance of its foreign policy. Both rhetoric and action suggest that Beijing's core objectives are, in rough order of importance, preserving a peaceful international environment; avoiding encirclement or isolation; securing access to materials and markets; and promoting the "democratization of international relations" (*guoji guanxi de minzhuhua*)—or multipolarity.

Its rethinking on how to achieve these objectives has been shaped by the challenges that, by the mid-1990s, had come to impinge on its ability to achieve those objectives using relatively passive measures. Beijing's response, as suggested earlier, includes a number of elements. These include developing and engaging with regional multinational organizations, deepening bilateral ties with designated strategic partners, securing equity stakes in resource production, and improving China's overall image as a constructive member of international society. In one sense these new, or newly important, elements of Chinese foreign policy are means to larger ends. Having been identified as critical to Beijing's success, however, they also represent a new set of intermediate objectives. They are now, therefore, both important means and ends in Chinese foreign policy.

Neither Beijing's rhetoric nor practice has ever strayed from its national interests, but changes in the international environment, and in China itself, have shifted its definition of the national interest. As with most states, Beijing's foreign policy contains contradictory strands. Borrowing the lexicon of international relations theory, these include both realist and liberal elements.[57] Emphasis on relations with strategic partners is generally consistent with a traditional, realist approach to foreign policy.[58] Multilateralism, confidence building, and some aspects of Chinese rhetoric, on the other hand, suggest a different understanding—one that seeks to defeat security dilemmas through systematically improved diplomatic and economic relations. Beijing's strategic culture remains large realist, but its center of gravity has shifted somewhat to take on liberal aspects.

Effectiveness

Since the late 1990s, Beijing has arguably been more successful internationally than at any other point in the history of the People's Republic. Given its spotty historical record, this may be setting the bar low, and certainly large questions remain about the efficacy of some of its efforts. Nevertheless, China has, at least thus far, successfully dealt with a daunting set of challenges that, with a less adept approach, might have damaged its economic prospects and left it isolated on the world stage.

Not least among its achievements, Beijing has avoided a broad-based balancing reaction from its neighbors, despite having added substantially to its hard power. It has established a leadership role for itself in Central and East Asia, though clearly most regional states want other major powers sharing leadership. Beijing has also blazed new trails in other regions, establishing strong relationships with a series of states in far away corners of the world. Economic frictions remain, but Beijing's economic diplomacy has enabled an explosion of trade, with inflation-adjusted trade growth averaging 15 percent between 1998 and 2005, and robust expansion of both inward and outward FDI. Beijing has gained credibility in many parts of the world as a constructive member of international society. According to a recent survey of attitudes in twenty-three countries (twenty-one of which are democracies), China is seen as having a significantly more positive impact in the world than the United States and runs only slightly behind Britain.[59]

Beijing has been able to achieve its objectives despite spending a smaller percentage of its resources on foreign policy than during the 1960s and 1970s. Its foreign policy is smarter, rather than more muscular. The government can now draw on a deeper reservoir of assets, many of which it does not need to finance or manage directly. But while China has clearly capitalized on its enhanced "comprehensive national strength," greater economic or political power cannot fully explain its success. Indeed, the country's rising national power makes its foreign policy requirements more demanding, even as it provides it with an expanded tool kit. Only flexible and sometimes creative policy adjustments—including multilateral engagement and confidence building, relatively generous terms in its trade and investment, and intensified bilateral efforts with selected partners—have enabled Beijing to turn its new power into a net diplomatic asset rather than a liability.

It would, however, be as inappropriate to exaggerate China's diplomatic success as to underrate it. In the regions around China's periphery, security concerns have been mitigated, but uncertainties over China's future directions and intentions continue to encourage hedging. Similarly, trade frictions have been managed, but have not disappeared. Most importantly, the international goodwill toward China reflected in recent surveys (including the one just mentioned) is fragile and could change quickly if China attempts to actually influence the behavior and policies of its partners—a subject addressed further below in the discussion of the range and nature of China's influence in the developing world and its impact on U.S. interests.

International Impact

More important than questions about the effectiveness of China's foreign policy in achieving its own national goals are questions about its impact on the world. What are the implications of its new foreign policy for international stability and peace, particularly in regions abutting China? What impact does it have on the United States' ability to achieve its objectives in the developing world? And what does it mean for the continued development and strengthening of international norms on issues related to human rights, corporate governance, democratic governance, and sovereignty? The answer to each of these questions is likely very different.

China's new diplomacy has had a mixed impact on the prospects for regional stability in Asia. Certainly, the concerns have changed. During the early 1990s, many U.S.-Asia observers worried that Asia was "ripe for rivalry," and that China might openly challenge the status quo or even employ military force to secure irredentist objectives in the South China Sea or elsewhere.[60] Concerns about possible war across the Taiwan Strait, and about Beijing's tensions with Japan, remain high. But the more generalized apprehension about China's territorial ambition has been considerably mitigated by its current emphasis on multilateralism, confidence building, and the resolution of most of its disputes over border demarcation.

Globally, China's investments and trade have had a net beneficial effect on the economic prospects for many developing states. Its investments in places like Argentina and South Africa provide a welcome injection of funds into parts of the world that might otherwise be neglected. Within individual countries, there are winners and losers that result from economic interaction with China, but trade and investment bring net benefits to its partners. As Beijing's economic interests extend progressively farther from home, its interest in stability-enhancing political efforts in more distant areas is also expanding.

The impact of China's new diplomacy on U.S. interests is more complicated.[61] On the positive side, China's growing interest in political stability in the developing world helps anchor a stable environment in which the United States can work to achieve other objectives—not least those related to winning the war on terror. This is particularly true in Central and Southeast Asia. Some Chinese scholars and government researchers have expressed a general interest in systematic cooperation in the problem of failed states.[62] With the United States suffering from "donor fatigue" and heavily tied down in the Middle East, China's expanded role in UN peacekeeping activities offers potentially significant assistance in these and other states. China's relative openness to imports and investment also provides a model for other developing states that is, despite some significant problem areas, largely salutary.[63]

Beijing's new approach, however, also poses potential new challenges. Unlike those of the mid-1990s, present concerns revolve as much around the attractiveness of Beijing's smiling visage as around its military behavior. China may prove so adept diplomatically, and have so much to offer, that its influence may increase at the expense of the United States. Should the trend continue, the United States may find it harder to mobilize nations on projects of interest. The Shanghai Cooperation Organization's October 2005 call for a timetable for the withdrawal of U.S. forces from Afghanistan indicates that, in certain cases, Beijing is willing to use its influence to try to constrain U.S. power.

For the most part, however, China's growing economic clout and the goodwill toward it generate more "defensive" benefits for it than "offensive" influence. China's new weight almost certainly increases the willingness of its partners to forgo efforts to hurt China (whether they be UN sanctions on human rights or military efforts that might appear directed at containing China), but these same partners would be far more reluctant to agree to more ambitious designs (e.g., measures that would limit U.S. influence or directly damage its interests). It is, in other words, more focused on counter-containment than comprehensive balancing against the United States. Indeed, part of Beijing's success in generating goodwill is explained by the fact that it generally only asks partners to modify their international behavior in ways that advance China's defensive interests.

Beijing's efforts to contain Taiwan—something Chinese leaders would define within this defensive context but which those in Taipei certainly would not—is, of course, an important, if partial, exception. Despite Beijing's interests in regional and global stability, cross-Strait dynamics could certainly prove disruptive. Other exceptions, such as the SCO's 2005 call for a U.S. timetable for withdrawal from Afghanistan, may be more apparent than real. In the SCO case, Beijing did not initiate (or by most reports lead) the discussion that resulted in the call for withdrawal, but rather lent its vote to a position strongly advanced by its Central Asian partners. Beijing's focus on achieving defensive political ends is not a function of virtue (any more than the behavior of most states), but rather of a realistic appraisal of the limitations of its own influence. This understanding, in turn, is informed by the occasional failure—the pushback encountered, for example, when Beijing overreached in its effort to secure an exclusive approach to the first East Asian Summit.

We can expect the scope of Beijing's influence to expand somewhat over the next decade. This will be particularly true in areas where the U.S. presence and attention is minimal. (Barring a course change by the United States, this will be nowhere truer than in Africa.) But the growth of Chinese influence and ambition is likely to be limited by a variety of factors.

Indeed, around China's periphery, the growth of Chinese power is likely to encourage a stronger political and military embrace of the United States, even as local states look to cash in on the economic opportunities provided by China. According to a November 2005 survey by the U.S. State Department, a significant plurality in most Southeast Asian states saw China as their nations' "most important economic partner in five to ten years." But by an equally large margin, the populations in these states saw the United States as their most important security partner—a result that indicates strong uncertainty and wariness in local states about China's future political and military direction.[64]

Economic and social considerations in China itself will also likely serve to limit its goals and keep its leaders focused on economic and limited (generally defensive) political goals, rather than more ambitious ones. Despite the growth of China's economy, its social and economic weaknesses are legion.

Chinese industrial innovative capacity is limited, and its export growth, particularly in the high-tech sectors, is heavily dependent on FDI from the United States and its allies—and by some measures is becoming more dependent on them.[65] Among China's many daunting social challenges, none is larger than the employment problem. China must find new jobs for an enormous number of people every year, including new entrants to the workforce (roughly 12 million a year), individuals laid off from state-owned enterprises (6 million), and migrants from the countryside (6 million).

Perhaps the most challenging aspect of China's rise in the near and midterm, therefore, will be managing its impact on evolving global norms, rather than on stability (where its interests largely, but not entirely, coincide with those of the United States).[66] Specifically, Beijing's advocacy of traditional conceptions of sovereignty, combined with its drive for equity stakes in natural resource extraction in the developing world, threaten progress in the area of corporate responsibility and transparency, democratic governance, and respect for human rights. Chinese oil companies have, with Beijing's backing, increasingly exploited opportunities in states where human rights or domestic political politics has made the regimes into international pariahs.

Chinese firms are not alone, especially in the ranks of developing states, in seeking space in isolated corners to ply their trade. In Sudan, for example, India's ONGC Videsh and Malaysia's Peremba are—together with China's China National Petroleum Corporation (CNPC)—major players.[67] Depending on political and economic circumstances, European states and Japan have also proven willing to exploit opportunities with questionable regimes.[68] Western (including U.S.) firms, especially in the extractive industries, have been accused of ignoring, or even participating in, abuses in host countries in any number of historical cases.[69] But while the idea of corporate responsibility has begun to have a larger impact on the practices of Western firms, it has yet to have a similar impact on Chinese companies or on state-owned firms from several other developing states. Given the speed with which these companies and their governmental patrons are emerging as important global investors, the potential impact on both recipient countries and the international behavior of rivals is significant.

It is not inconceivable that China's resource diplomacy may evolve to accommodate global norms. Chinese leaders have proven capable of reevaluating their priorities and interests on a broad range of issues. The government may ultimately decide that protecting its international reputation outweighs the potential benefits of certain types of relationships with pariah states. At the corporate level, reputation may begin to weigh more heavily on decision makers as their firms are listed on foreign bourses—subjecting Chinese companies to the same kinds of pressures to which Western corporations are currently subjected. However, there is nothing inevitable about all of this. Certainly in the short and medium term, China's growing dependence on imported energy—combined with high energy prices and instability in the Middle East—will produce strong countervailing pressures, as they will on other governments.

Conclusion

China's foreign policy remains a work in progress. It shed its ideological clothing, becoming pragmatic and largely economically oriented after 1978. The limits of pragmatism, at least when unaccompanied by proactive diplomatic support, became apparent during the mid-1990s, and Beijing has since made significant adjustments. Thus far, China's new foreign policy has served its national interests well. To be sure, questions remain about how much actual influence it can exercise, tensions remain in its priorities, and (like any state) it sometimes blunders. But the nation's softer face, and the multilateral economic and political engagement that has accompanied it, has enabled Beijing to improve its relations with nearby states even as its hard power has increased rapidly. Globally, deep engagement with key partners has enabled trade outlets and resource acquisitions to keep pace with and support a $1.8 trillion economy growing at a sustained rate of 9 percent.

Beijing will continue to present both challenges and opportunities for the United States. The magnitude of these challenges will depend on how China's foreign policy evolves over time. Many of the variables in Beijing's future direction will be largely exogenous to U.S. foreign policy, but no single country can have a more profound impact on China's view of its interests than the United States.

Washington should be alert for areas in which it can work with Beijing in the developing world. U.S. and Chinese interests in the developing world sometimes largely overlap—particularly in buttressing stability, preventing state failures, and fighting terrorism and the spread of nuclear weapons. At the same time, Washington should continue to encourage Beijing to be a responsible member of international society and support global norms on human rights and corporate responsibility. U.S. leaders have every right to challenge Beijing when it undermines, rather than supports, these norms, but they should also remember that abiding by emerging global norms will be more palatable to Beijing if it also has some say in shaping the rules of the road and only if it believes the United States (and others) will apply norms and standards uniformly.

Finally, American diplomatic, economic, and military policy should seek to ensure that the United States remains a respected and important player in all regions of the world. This will require improvements in America's ability to multitask: to fight the war on terror and achieve other immediate security interests, while still addressing concerns in Asia and elsewhere that may loom as large or larger for local governments. It will require an ability to keep an eye on—and hand in—regions like Sub-Saharan Africa that may not be on the front pages but where U.S. long-term interests are nevertheless engaged.

Notes

This chapter is the product of extended conversations with the volume's authors and editors. It is built on their observations and ideas, and has benefited immeasurably from their comments and suggestions. Although I have tried to represent the views of the group, given the range of ideas held, not every contributor may agree with all of its conclusions. In addition to the authors and editors, I have benefited greatly from comments by and discussions with Evan Medeiros, George Gilboy, Christopher Twomey, and Phillip Saunders. Of course, any factual or analytical errors are my responsibility alone.

1. A number of authors have commented on the dramatic changes in Chinese foreign policy over the last several years. See Phillip Saunders, *China's Global Activism: Strategy, Drivers, and Tools* (Washington, D.C.: National Defense University Press, June 2006); Avery Goldstein, *Rising to the Challenge: China's Grand Strategy and International Security* (Stanford, CA: Stanford University Press, 2005); Evan S. Medeiros and M. Taylor Fravel, "China's New Diplomacy," *Foreign Affairs* (November/December 2003); Rosemary Foot, "Chinese Power and the Idea of a Responsible State," *China Journal* (January 2001); and David Shambaugh, "China Engages Asia: Reshaping the Regional Order," *International Security* (Winter 2004/2005).

2. For a review essay on China's foreign policy during and after the Cultural Revolution, see Thomas W. Robinson, "Chinese Foreign Policy from the 1940s to the 1990s," in *Chinese Foreign Policy: Theory and Practice,* ed. Thomas W. Robinson and David Shambaugh (New York: Oxford University Press, 1994).

3. The balance between ideology and realism in China's foreign policy shifted over time. Beijing's approach was relatively pragmatic and state-centered during the 1950s. It was more ideological and revolutionary during the first phases of the Cultural Revolution, and it edged once again toward realism as China's competition with the Soviet Union heated up during the 1970s. Regardless of some variation, though, Beijing's foreign policy retained a strong ideological tint until 1978.

4. A full text search of the *Jiefangjun Bao* shows the use of the phrase "developing states" (*fazhanzhong guojia*) becoming more common than "third world" (*disan shijie*) in 1987.

5. Nina Halpern, "Social Scientists as Policy Advisers in Post-Mao China: Explaining the Pattern of Advice," *Australian Journal of Chinese Affairs,* no. 19/20 (January–July 1988). For a more recent review of individual foreign policy think-tanks, see Bonnie S. Glaser and Phillip C. Saunders, "Chinese Civilian Foreign Policy Research Institutes: Evolving Roles and Increasing Influence," *China Quarterly* (2002); and David Shambaugh, "China's International Relations Think Tanks: Evolving Structure and Process, *China Quarterly* (2002).

6. For more on the challenges facing Chinese foreign policy during the mid-1990s, see Goldstein, *Rising to the Challenge,* 102–117.

7. Gerald Segal, "China and Africa," *Annals of the American Academy of Political and Social Science* (January 1992).

8. According to guidelines established by Zhao Ziyang in 1983, projects were to "achieve good economic results with less investment, shorter construction cycles and quicker returns." George T. Yu, "Africa and Chinese Foreign Policy," *Asian Survey* (August 1988): 858.

9. Richard A. Bitzinger, "Arms to Go: Chinese Arms Sales to the Third World," *International Security* (Autumn 1992).

10. Unless otherwise noted, all trade statistics in this chapter are from the International Monetary Fund's *Direction of Trade* (DOT) statistics. Also unless otherwise noted, the figures include combined totals for Hong Kong and the mainland, minus trade with one another (in other words, the figures treat the two as a single unit).

11. Between 1980 and 1989, China established formal diplomatic relations with eight Latin American states, bringing the total to twenty. Ten Latin American presidents, eight prime ministers or vice prime ministers, and thirty legislative delegations visited Beijing. Frank O. Mora, "Sino-Latin American Relations: Sources and Consequences, 1977–1997," *Journal of Interamerican Studies and World Affairs* (Summer 1999): 97. Total two-way trade with Latin America grew by over 600 percent, though at $4.9 billion in 1989, it remained only 12 percent as large as China's $35 billion in trade with developing East Asia.

12. At the ASEAN post-ministerial conference in 1992, China's Foreign Minister Qian Qichen suggested that ASEAN and China sign a partnership agreement, but the proposal went nowhere. *Joint Communiqué*, 25th ASEAN Ministerial Meeting, Manila, Philippines (July 21–22, 1992), available at www.aseansec.org/3667.htm.

13. In April 1997, Admiral Joseph Prueher, the commander-in-chief of the U.S. Pacific Command, became the highest ranking officer to visit Hanoi since the Vietnam War. "Now Vietnam Needs America to Ward off China," *Christian Science Monitor* (April 17, 1997).

14. Vietnam was admitted in 1995, Laos and Myanmar in 1997, and Cambodia in 1999.

15. As far back as 1992, a draft copy of the Pentagon's biennial Defense Planning Guidance had stipulated that a key U.S. goal would be maintaining America's sole superpower status and "convincing potential competitors that they need not aspire to a greater role." While China took note, there is little evidence that it was alarmed by this declaration of American military supremacy—or by its basis in fact. Patrick E. Tyler, "U.S. Strategy Plan Calls for Insuring no Rivals Develop," *New York Times* (March 8, 1992).

16. Full-text searches of both *Renmin Ribao* and *Jiefangjun Bao* show a spike in mentions of "hegemonism" (*baquan zhuyi*) in 1997 that has only gradually subsided since.

17. By 1995, imports from these states already exceeded exports to them by 51 percent, and by 2003 the figure had risen to 83 percent.

18. A report by the China Institute of Contemporary International Relations (CICIR) on China's relationship with Latin America notes the local perception that the industrial structures of the two are more competitive than complementary, as well as noting a general sensitivity to "economic colonialism." The CICIR report claims that "anti-dumping duties as high as 500–1800 percent" have been imposed on some 4,500 kinds of goods. "Report on China's Latin America Policy," *Contemporary International Relations* (April 2004): 29.

19. "The Hungry Dragon," *Economist* (February 19, 2004); "Trade with China and India: Fears and Hopes," *Business Recorder* (February 9, 2005).

20. "Chinese Oil Imports to Meet 50 Percent of Demand by 2010: Report," AFP (February 16, 2005).

21. The idea was further developed and became enshrined in China's larger diplomatic doctrine during the 1955 Asian-African Conference in Bandung. The five principles include: mutual respect for sovereignty and territorial integrity, mutual non-aggression, non-interference in each other's internal affairs, equality and mutual benefit, and peaceful coexistence.

22. Zhu Mingquan, "Beyond Westphalia and New Security Concepts," *Hong Kong Baptist University Department of International Studies Working Paper* (March 2005). The specific content of the new security concept differs from one presentation to the next. Some versions of the concept, especially those outlined by military officers or analysts, contain relatively more hard-edged realist elements. The ideas listed here, however, are consistent with most versions, especially those outlined by senior civilian leaders. See, for example, comments by Qian Qichen during a speech celebrating ASEAN's thirtieth anniversary in December 1997. "Qian Qichen Zai Jilongpo Chanshu Xin Anquanguan" (Qian Qichen elaborates on the new security concept in Kuala Lumpur), *Jiefangjun Bao* (December 16, 1997). They are also broadly consistent with comments made by Hu Jintao in Indonesia in July 2000 and in France in November 2001.

23. This discussion began in 2002, largely in response to concerns in the United States and Southeast Asia that China's rise might be highly destabilizing. By early 2004, the term was in wide use by senior leaders, including Premier Wen Jiabao and President Hu Jintao. Use of the term by most senior civilian officials ended in April 2004, a victim of criticism that, on the one hand, China had not yet "risen" and, on the other, that Beijing should not rule out the use of force against Taiwan. Nevertheless, the idea continues to be debated in think tanks, and the phrase "peaceful emergence" continues in use by Foreign Ministry and other officials. See Robert L. Suettinger, "The Rise and Descent of 'Peaceful Rise,'" *China Leadership Monitor,* no. 12 (Fall 2004); Evan S. Medeiros, "China Debates Its 'Peaceful Rise' Strategy," *YaleGlobal Online* (June 22, 2004), available at http://yaleglobal.yale.edu/article.print?id=4118; "Ambassador Wu Hongbo Delivered Welcome Address at the Conference on China's Peaceful Emergence in East Asia" (February 25, 2005).

24. See, for example, Zhang Chengzhi and Yang Zidi, "'Shanghai Jingshen': Kaichuang Guoji Heping Hezuo Xin Geju de Qizhi" (The Shanghai [five] spirit: planting the banner of peaceful international cooperation), *Jiefangjun Bao* (June 18, 2001). Jia Qingguo, "The Success of the Shanghai Five: Interests, Norms and Pragmatism," available at www.ndu.edu/inss/ symposia/pacific2001/jiafinal.html.

25. At the time the code of conduct was signed, the most intense outstanding disagreements were between the Philippines, Vietnam, and China. In March 2005, however, these three countries signed an agreement to conduct joint exploration for oil in disputed territory. See "China, Vietnam, and Philippines Sign Joint Oil Exploration Accord for South China Sea," *World Markets Analysis* (March 15, 2005); Karen Teo, "Petrochina to Explore Near Spratlys for Oil and Gas," *The Standard* (March 26, 2005).

26. Although it could theoretically reverse its position later, the costs of such a reversal will only grow as its trade with regional states grows.

27. For more on the maneuvering surrounding China's membership in these groups, see chapter 4 by Chung-chian Teng in this volume.

28. "Full Text of Jiang Zemin's Report at 16th Party Congress on November 8, 2002," available at www.chinaembassy.org.in/eng/zyjh/t61448.htm.

29. Word searches in full-text databases of two Chinese newspapers show that mentions of "developing states" and "the third world" (combined) doubled between 1990 and 2000. *Jiefangjun Bao*, full-text CD ROM; *Renmin Ribao*, full-text CD ROM. Mentions tripled between 1980 and 2000.

30. Yan Shiliang, "Lun Lengzhan Hou Fazhan Zhong Guojia de Diwei yu Zuoyong" (Discussing the importance and function of developing states after the Cold War), *Guoji Wenti Yanjiu* (International studies) (January 20, 2000).

31. There are several levels of partnership in China's diplomatic lexicon, of which "strategic" partnerships are at the top. The discussion below is restricted to this latter, more selective category.

32. Some media reports also reference a strategic partnership with Egypt (1999), though this appears to be dormant. Strategic partnerships have also been concluded with several advanced industrial states: Russia (1996), France (1997), the European Union (2003), the United Kingdom (2004), Italy (2004), Canada (2005), Portugal (2005), and Greece (2006). This list was compiled from Chinese media reports and announcement on the Chinese Ministry of Foreign Affairs Web site and from data supplied in Evan S. Medeiros, *China's International Behavior* (Santa Monica: Rand Corporation, forthcoming).

33. This counter-containment incentive in Chinese foreign policy more generally is a central argument made by Evan Medeiros in Medeiros, *China's International Behavior.*

34. Ministry of Commerce of the People's Republic of China, report on 2005 outward FDI (in Chinese), http://hzs.mofcom.gov.cn/aarticle/date/200609/2006090 3095437.html; *China Statistical Yearbook, 2005,* 643 and 650.

35. The package included $5 billion in oil and gas projects; $8 billion in railway systems; $6 billion in housing construction; and $750 million in communications projects. These were not definitive commitments, but came with a firm "offer of finance" by the Chinese government. See "Kirchner Welcomes Help from the East," *Brazil Report* (November 23, 2004).

36. On the Chinese debate over energy security, see George Gilboy, "Crossing Energy Security Policy After September 11: Crossing the River While the Stones Are Moving," *CERA Private Report* (February 2002); Erica S. Downs, "The Chinese Energy Security Debate," *China Quarterly* (March 2004).

37. On some of these investments, see Brian Bremner and Dexter Roberts, "The Great Oil Hunt," *Business Week* (November 15, 2004).

38. "Argentina Oil Deal with China Gives Angola a Key Role," Associated Press (November 22, 2004). "China Sonangol, Intl, Argentine Enarsa to Cooperate in Offshore Exploration," *Latin America News Digest* (December 3, 2004). More generally, it is significant that despite high oil prices, profit-oriented Western multinationals are not, for the most part, expanding their operations or investments anywhere near as fast as Chinese oil companies. "Oil Companies Reap Cash Windfall from Price Surge," *Wall Street Journal* (March 18, 2005).

39. "China, ASEAN Study Free Market for Farm Produce," *Asia Pulse* (December 20, 2002). The agreement also requires China and the six more economically developed ASEAN states to lower their tariff barriers by 2010, but allows the four weaker economies of Cambodia, Laos, Myanmar, and Vietnam to maintain theirs until 2015. "ASEAN, China Ink Pact to Deepen Economic Cooperation," Kyodo (November 11, 2002).

40. "China Says Reaches 'Understanding' with Brazil on Soy Trade," *AFX* (June 23, 2004). Beijing's flexibility has not been restricted to the agricultural sector. In June 2004, China hosted a delegation of 450 Brazilian businessmen to explore areas where they might find markets for industrial products. "Trade-Argentina: President Heads to China to Boost Ties," IPS-Inter Press Service (June 24, 2004).

41. For the most comprehensive source on China's foreign aid thinking, as well as facts and figures, see Michael A. Glosny, "Meeting the Development Challenge in the 21st Century: American and Chinese Perspectives on Foreign Aid," *National Committee on U.S.-China Relations, China Policy Series* (2006).

42. *China Statistical Yearbook 2003* (Beijing: National Bureau of Statistics of China, 2003), 292*; China Statistical Yearbook 2004* (Beijing: National Bureau of Statistics, 2004), 308; and *China Statistical Yearbook 2005* (Beijing: National Bureau of Statistics, 2005). Figures are provided in yuan: 5.00 billion in 2002 and 5.22 billion in 2003.

43. See Han Chiao, "China's Annual Foreign Aid Reaches 15 Billion Yuan," *Cheng Ming* (August 1, 2004).

44. During the financial crisis, China gave $300 million in assistance to Indonesia. Following the tsunami, China contributed $100 million (including $83 million in official assistance) to aid Indonesia, Thailand, and India. The United States contributed around $1 billion, while Japan's contribution was about half that. As a percentage of GDP, the contribution of the three was more roughly equal, but the difference in absolute numbers highlights the continued importance of economic bulk.

45. Richard F. Grimmett, "Conventional Arms Transfers to Developing Nations, 1996–2003," *CRS Report for Congress* (August 26, 2004): 43–44.

46. Together with Russia, Iran, and Belarus, China is one of the primary sources of military hardware to Sudan, a country that not coincidentally accounts for one-tenth of China's total oil imports. "Sudan: Arms Trade Fuelling Human Rights Abuse in Darfur," Amnesty International Press Release (November 11, 2004). Stockholm International Peace Research Institute, "Arms Transfers to Sudan, 1994–2004," available at www.sipri.org/contents/armstrad/atsud_data.html.

47. "New Tour Destinations, New Prosperous Outlook," *Business Daily Update* (July 6, 2004).

48. "China Sees Rising Influx of Foreign Students," Xinhua News Agency (July 11, 2006); "Ministry Reports Record Number of Foreign Students Studying in China in 2004," Xinhua News Agency (May 21, 2005).

49. "NPC Deputy Calls for Promoting Chinese," Financial Times Global News *Wire* (March 10, 2006).

50. Since 1978, the PRC has joined the IMF (1980), the World Bank (1981), the IAEA (1984), and the WTO (2001). It gained China's seat on the UN and associated organizations in 1971.

51. China increased its contribution from $6.2 billion to $8.4 billion, lifting its voting power within the IMF from 2.2 percent to 3.0 percent. See "At a Glance—China and the IMF," *International Monetary Fund* (September 1, 2004), available at www.imf.org/external/country/chn/rr/glance.htm.

52. "First UN Agency's Headquarters Established in China," Xinhua (November 19, 2003). The headquarters is for the Asian and Pacific Center for Agricultural Engineering and Machinery. China previously had branch offices of various agencies on its soil.

53. "China Boosts Support for IAEA Development and Security Initiatives," IAEA Press Release, September 2004.

54. "China the Largest UNSC Contributor to Peacekeeping Missions," *People's Daily* (September 29, 2006).

55. "China Filling Void Left by West in U.N. Peacekeeping," *Washington Post* (November 24, 2006).

56. For the critical view, see Bill Gertz, "China Will Send Troops to Haiti," *The Washington Times* (September 6, 2004). Gertz worries that China might use its presence in Haiti to encourage Port au Prince to switch its recognition from Taiwan to the mainland.

57. "Realist" and "liberal" refer to ideas consistent with two schools of thought on international relations theory. "Realist" does not, here, imply a more "realistic" approach to foreign policy. And "liberal," does not suggest one that is more respectful of individual rights. Rather, realist refers to an approach that emphasizes state power, especially military power, as the key to security. Liberal approaches are those that posit security competition can be mitigated—and therefore security enhanced—by confidence building and other cooperative efforts among states. For a relatively brief discussion of these ideas, see Paul R. Viotti and Mark Kauppi, *International Relations Theory,* 3rd ed. (New York: Macmillan, 1999).

58. Beijing would argue that since its partnerships are not alliances and not exclusive, they do not create or reinforce zero-sum dynamics in international relations. Nevertheless, the emphasis on special, bilateral, and strategic ties represents a largely traditional, or realist, approach to international relations.

59. Program on International Policy Attitudes (PIPA), "In 20 of 23 Countries Polled Citizens Want Europe to Be More Influential than U.S." (April 6, 2005). The 23 countries surveyed included twenty-one democracies and were selected from Europe, Asia, Latin America, and the Middle East. PIPA is a program affiliated with the University of Maryland.

60. Aaron L. Friedberg, "Ripe for Rivalry: Prospects for Peace in a Multipolar Asia," *International Security* (Winter 1993/1994); Denny Roy, "Hegemon on the Horizon? China's Threat to East Asian Security," *International Security* (Summer 1994).

61. For an effort to catalog common and divergent interests between China and the United States and evaluate the balance between the two, see Shambaugh, "China Engages Asia: Reshaping the Regional Order," *International Security*, vol. 29, no. 3 (Winter 2004/05).

62. Banning Garrett and Jonathan Adams, "U.S.-China Cooperation on the Problem of Failing States and Transnational Threats," *United States Institute of Peace Special Report 126* (September 2004).

63. Real concerns abound, especially in the area of intellectual property rights. But unlike Japan and South Korea, especially during their takeoff periods, China's development is being accomplished on the back of relatively free trade and inward investment. With an economy somewhat more than one-third the size of Japan's, the total value of its trade has already surpassed that of Japan. Unlike its neighbor, China's trade is generally balanced and sometimes in deficit. Foreign FDI in China is roughly eight times greater than Japan's inward FDI.

64 "Asian Views of China," Department of State, November 9, 2005.

65.On the limitations of Chinese industrial innovative capacity, see George J. Gilboy, "The Myth Behind China's Miracle," *Foreign Affairs* (July/August 2004). Foreign-invested firms account for more than 50 percent of China's total exports. They account for a far higher percentage (between 74 and 92 percent, depending on the sector) of high-tech exports—and the percentage has grown rapidly since the early 1990s.

66. The distinction between China's impact on stability, U.S. interests, and global norms is at least partially artificial. The United States has interests in stability and norms. Nevertheless, it is useful to think about each in turn, since the three are not always synonymous.

67. On Sudan, see "Country Analysis Brief," *Energy Information Administration: Official Energy Statistics from the U.S. Government*, available at www.eia.doe.gov/emeu/cabs/sudan.html. Indian energy companies have also competed against Chinese energy companies in Burma, Angola, and Venezuela, among other places. There are structural reasons, apart from tighter energy markets, for the scramble among national oil companies for equity stakes. Having matured in their home markets, companies from a number of developing countries are increasingly looking for easy pickings abroad where U.S. and Western firms cannot, for political reasons, compete.

68. Landing its first large contracts in 2004, China was a latecomer to Iran's oilfields, joining established European companies, as well as Japanese firms that arrived several years earlier. Prior to 2004, China had previously undertaken only small investment deals with Tehran relative to those of several European states.

69. For a good synopsis of the evolution of the sensitivity of Western firms to moral pressures, see Marina Ottaway, "Reluctant Missionaries," *Foreign Policy* (July/August 2001).

IV

Appendices

Appendix I. **Trade Statistics for China Plus Hong Kong** (in $U.S. millions)

	1961	1978	1979	1980	1989	1990	1991	1992	1993	1994	1995
Imports from World	1,508	21,975	29,481	36,820	93,077	90,988	108,408	138,573	179,343	206,776	246,258
Exports to World	918	18,555	24,956	32,027	84,148	96,185	110,191	130,389	159,001	187,982	226,168
Total World	2,426	40,530	54,436	68,847	177,225	187,173	218,598	268,962	338,344	394,758	472,426
Total (in constant 2004 $US)	15,327	117,425	141,639	157831	269,982	270,520	303,181	362,131	442,305	503,170	585,573
Annual Growth						0.20	12.07	19.44	22.14	13.76	16.38
Imports from Ind.[a]	1,123	15,080	20,569	26,175	59,819	58,209	68,099	84,556	105,776	122,732	143,162
Exports to Ind.	581	11,599	15,991	20,424	58,003	65,630	75,691	89,444	113,260	134,578	157,692
Total Ind.	1,704	26,679	36,560	46,599	117,822	123,839	143,790	174,000	219,036	257,310	300,854
Total (2004 $US)	10,763	77,295	95,125	106826	179,489	178,983	199,427	234,274	286,338	327,975	372,909
Imports from Non-Ind.[b]	385	6,895	8,912	10,645	33,258	32,779	40,308	54,016	73,567	84,044	103,096
Exports to Non-Ind.	337	6,956	8,965	11,604	26,145	30,555	34,500	40,945	45,741	53,404	68,476
Total Non-Ind.	722	13,851	17,877	22,249	59,403	63,33	74,808	94,962	119,308	137,448	171,573
Total (2004 $US)	4,564	40,131	46,514	51,005	90,494	91,536	103,754	127,857	155,967	175,195	212,665
% of World Trade	30	34	33	32	34	34	34	35	35	35	36
Imports from Africa	49	360	438	599	964	867	934	1,035	1,402	1,377	2,109
Exports to Africa	66	510	533	1,242	1,729	2,573	2,455	3,017	3,043	3,169	4,207
Total with Africa	115	870	971	1,841	2,693	3,440	3,389	4,052	4,445	4,546	6,316
Total (2004 $US)	727	2,520	2,526	4,220	4,102	4,972	4,700	5,456	5,811	5,794	7,829
% of Total Trade	4.7	2.1	1.8	2.7	1.5	1.8	1.6	1.5	1.3	1.2	1.3
% of Trade w/Non-Ind.	15.9	6.3	5.4	8.3	4.5	5.4	4.5	4.3	3.7	3.3	3.7

Imports from Asia	227	3,204	4,219	6,256	20,428	24,405	32,168	41,438	56,947	67,774	84,008
Exports to Asia	174	2,785	3,583	5,523	14,917	18,347	22,041	24,377	26,930	33,467	44,153
Total with Asia	400	5,989	7,802	11,779	35,345	42,752	54,209	65,815	83,876	101,241	128,161
Total (2004 $US)	2,529	17,351	20,300	27,003	53,844	61,789	75,185	88,614	109,649	129,045	158,856
% of China's Total Trade	16.5	14.8	14.3	17.1	19.9	22.8	24.8	24.5	24.8	25.6	27.1
% of Trade w/ Non-Ind.	55.4	43.2	43.6	52.9	59.5	67.5	72.5	69.3	70.3	73.7	74.7
Imports from Middle East	40	528	481	638	1,304	1,160	1,529	2,097	3,086	3,304	3,990
Exports to Middle East	35	850	1,176	1,779	2,648	2,679	3,182	3,987	4,870	5,378	6,265
Total with Middle East	75	1,378	1,657	2,417	3,952	3,839	4,711	6,084	7,955	8,633	10,255
Total (2004 $US)	473	3,993	4,312	5,541	6,021	5,549	6,534	8,192	10,400	11,067	12,711
% of China's Total Trade	3.1	3.4	3.0	3.5	2.2	2.1	2.2	2.3	2.4	2.2	2.2
% of Trade w/ Non-Ind.	10.4	9.9	9.3	10.9	6.7	6.1	6.3	6.4	6.7	6.3	6.0
Imports from W. Hemisphere	17	104	119	207	2,892	1,906	2,115	2,559	2,736	3,048	3,964
Exports to W. Hem.	20	263	543	1,003	1,517	2,178	2,854	3,902	5,125	6,574	7,907
Total with W. Hem.	37	366	662	1,210	4,409	4,084	4,969	6,461	7,861	9,622	11,870
Total (2004 $US)	236	1,061	1,723	2,775	6,716	5,903	6,892	8,699	10,276	12,265	14,713
% of Total Trade	1.5	0.9	1.2	1.8	2.5	2.2	2.3	2.4	2.3	2.4	2.5
% of Trade w/Non-Ind.	5.2	2.6	3.7	5.4	7.4	6.4	6.6	6.8	6.6	7.0	6.9
Other Non-Ind. States in Non-Ind. Total	13.1	37.9	38.0	22.5	21.9	14.6	10.1	13.2	12.7	9.7	8.7

Source: International Monetary Fund Direction of Trade Statistics, www.imf.org.
[a] Ind.=Industrialized states
[b] Non-ind.=Non-industrialized states

(continued)

Appendix I. *(continued)*

	1996	1997	1998	1999	2000	2001	2002	2003	2004	2005
Imports from World	255,600	264,883	243,040	259,883	324,257	347,820	400,261	532,579	702,524	812,275
Exports to World	234,396	259,209	256,576	272,034	335,021	337,839	386,757	487,977	602,378	793,859
Total World	489,997	524,092	499,617	531,917	659,278	685,659	787,018	1,020,556	1,304,902	1,606,134
Total (2004 $US)	589,932	616,829	579,004	603,116	723,215	731,344	826,391	1,047,733.9	1,304,902	1,553,501
Annual Growth	0.74	4.56	(6.13)	4.16	19.91	1.12	13.00	26.7	24.55	19.05
Imports from Ind.	143,231	145,325	134,525	141,716	167,696	178,884	191,729	243,278	308,586	329,217
Exports to Ind.	165,598	180,954	185,716	196,941	236,825	236,789	264,603	329,725	420,030	588,357
Total Ind.	308,829	326,278	320,241	338,657	404,521	415,673	456,332	573,003	728,616	917,574
Total (2004 $US)	371,816	384,012	371,126	383,987	443,752	443,369	479,161	588,262	728,616	887,505
Imports from Non-Ind.	112,369	119,558	108,515	118,167	156,561	168,936	208,532	289,301	393,938	483,058
Exports to Non-Ind.	68,798	78,256	70,861	75,093	98,196	101,050	122,154	158,252	182,348	205,502
Total Non-Ind.	181,167	197,814	179,376	193,260	254,757	269,986	330,686	447,553	576,286	688,560
Total (2004 $US)	218,117	232,817	207,878	219,129	279,463	287,975	347,229	459,472	576,286	665,996
% of China's World Trade	37	38	36	36	39	39	42	44	44	43
Imports from Africa	2,227	3,155	2,033	3,012	5,315	5,182	6,085	6,993	16,024	21,018
Exports to Africa	4,131	4,522	4,962	4,780	5,643	6,356	7,170	10,357	13,663	17,752
Total with Africa	6,358	7,677	6,995	7,792	10,958	11,538	13,255	17,350	29,687	38,770
Total (2004 $US)	7,655	9,035	8,106	8,835	12,021	12,30	13,918	17,812	29,687	37,500
% of China's Total Trade	1.3	1.5	1.4	1.5	1.7	1.7	1.7	1.7	2.3	2.4
% of China's Trade w/Non-Ind	3.5	3.9	3.9	4.0	4.3	4.3	4.0	3.9	5.2	5.6

Imports from Asia	90,120	95,670	89,329	95,428	119,426	124,355	152,109	203,871	268 919	316,778
Exports to Asia	44,476	49,807	41,443	46,644	61,082	61,442	74,733	94,088	94,603	157,298
Total with Asia	134,596	145,477	130,772	142,072	180,508	185,797	226,842	297,959	363,522	474,076
Total (2004 $US)	162,047	171,219	151,551	161,089	198,014	198,176	238,190	305,894	363,522	458,540
% of Total Trade	27.5	27.8	26.2	26.7	27.4	27.1	28.8	29.2	27.9	29.5
% of Trade w/Non-Ind	74.3	73.5	72.9	73.5	70.9	68.8	68.6	66.6	63.1	68.9
Imports from Middle East	4,656	5,608	4,311	4,970	10,868	11,086	11,896	17,424	25,772	36,237
Exports to Middle East	6,378	7,069	7,357	7,874	9,671	10,653	13,227	17,657	22,305	28,416
Total with Middle East	11,034	12,677	11,668	12,844	20,539	21,739	25,123	35,081	48,077	64,653
Total (2004 $US)	13,285	14,921	13,522	14,563	22,531	23,187	26,380	36,015	48,077	62,534
% of Total Trade	2.3	2.4	2.3	2.4	3.1	3.2	3.2	3.4	3.7	4.0
% of Trade w/Non-Ind.	6.1	6.4	6.5	6.6	8.1	8.1	7.6	7.8	8.3	9.4
Imports from W. Hemisphere	4,841	5,292	4,355	4,035	6,031	7,930	9,647	16,556	23,437	28,240
Exports to W. Hem.	7,634	9,654	9,989	8,894	11,729	12,334	13,252	14,659	21,570	26,968
Total with W. Hem.	12,475	14,946	14,344	12,929	17,760	20,264	22,899	31,215	45,007	55,208
Total (2004 $US)	15,020	17,590	16,623	14,660	19,482	21,614	24,045	32,046	45,007	53,399
% of Total Trade	2.5	2.9	2.9	2.4	2.7	3.0	2.9	3.1	3.4	3.4
% of Trade w/Non-Ind.	6.9	7.6	8.0	6.7	7.0	7.5	6.9	7.0	7.8	8.0
Other Non-Ind States in Non-Ind Total	9.2	8.6	8.7	9.1	9.8	11.4	12.9	14.7	15.6	8.1

Appendix II. **China's Oil Consumption from the Developing World, 2003–2005**

Region	Millions of Tons a Year			% of Total Oil Imports			% Change Y-on-Y	
	2003	2004	2005	2003	2004	2005	2003–2004	2004–2005
Africa	22.5	35.4	38.5	8.3	11.1	11.8	57.3	8.8
Middle East	51.8	62.8	67.4	19.1	19.7	20.6	21.2	7.3
South and Central America	2.3	4.1	5.3	0.8	1.3	1.6	78.3	29.3
Asia Pacific (excludes Japan)	34.0	40.0	30.3	12.5	12.5	9.3	17.6	–24.3
Total Oil Imports	271.7	318.9	327.3	—	—	—	17.4	2.6

Africa:
Territories on the north coast of Africa from Egypt to Western Sahara; territories on the west coast of Africa from Mauritania to Angola, including Cape Verde, and Chad; territories on the east coast of Africa from Sudan to the Republic of South Africa—also Botswana, Madagascar, Malawi, Namibia, Uganda, Zambia, and Zimbabwe.

Middle East:
Arabian Peninsula, Iran, Iraq, Israel, Jordan, Lebanon, and Syria.

South and Central America:
Caribbean (including Puerto Rico), Central and South America.

Asia Pacific:
Brunei, Cambodia, China, China Hong Kong SAR, Indonesia, Japan, Laos, Malaysia, Mongolia, North Korea, Philippines, Singapore, South Asia (Afghanistan, Bangladesh, India, Myanmar, Nepal, Pakistan, and Sri Lanka), South Korea, Taiwan, Thailand, Vietnam, Australia, New Zealand, Papua New Guinea, and Oceania.

Source: BP Statistical Review of World Energy, 2004–2006.

About the Contributors

Kurt M. Campbell is senior vice president and director of the International Security Program at the Center for Strategic and International Studies. Dr. Campbell served in several capacities in government, including Deputy Assistant Secretary of Defense for Asia and the Pacific in the Pentagon.

Joshua Eisenman is a Ph.D. student in political science at the University of California, Los Angeles, and a fellow in Asia studies at the American Foreign Policy Council in Washington, DC. He received his M.A. and B.A. in international relations from Johns Hopkins University's Paul H. Nitze School of Advanced International Studies and George Washington University, respectively. He has spoken publicly on China's strategy toward Africa at the National Defense University, the Council on Foreign Relations, and the Heritage Foundation, and has published articles on the topic in *Current History*, the *International Herald Tribune*, the *Straits Times*, the *South China Morning Post*, the *Asia Times*, and Hong Kong's *The Standard*. He has lived and traveled extensively in China and speaks Mandarin Chinese.

Michael A. Glosny is a Ph.D. candidate in political science at the Massachusetts Institute of Technology, a member of MIT's Security Studies Program, and a fellow at Harvard University's Belfer Center for Science and International Affairs. His research interests include international relations theory, Chinese foreign and security policy, and U.S. alliances and military policy in East Asia. His current project examines how rising powers manage their international environment during their rise, and involves extensive Chinese-language research and interviews. He has published articles and book chapters in *International Security, Asian Security, Strategic Asia 2003–4,* and the *China Policy Series* of the National Committee on U.S.-China Relations.

Eric Heginbotham is a political scientist at the RAND Corporation. He is a specialist in Chinese and Japanese foreign and security policy and has published articles in *International Security*, *Foreign Affairs*, *The National Interest*, and *Current History*, as well as chapters in several edited volumes. He holds a Ph.D. in political science from MIT, received his B.A. from Swarthmore College, and has spent twelve years in Asia.

Rollie Lal, Ph.D., is a political scientist at the RAND Corporation in Arlington, Virginia. She is an expert on Asian security and has written on a wide variety of issues related to India, China, Iran, Central Asia, North Africa, and political Islam. Dr. Lal is the author of *Understanding China and India: Security Implications for the United States and the World* (2006) and *Central Asia and its Asian Neighbors: Security and Commerce at the Crossroads* (2006), and coauthor of *The Muslim World after 9/11* (2004). She is also the author of articles in *Orbis*, the *Atlantic Monthly*, *Financial Times*, *International Herald Tribune*, and *Chicago Sun-Times*.

Mao Yufeng is a historian and specialist in Chinese-Middle Eastern relations and state-Muslim relations in China. She is currently completing a Ph.D. thesis on the topic at George Washington University.

Carola McGiffert is vice president and chief of staff at the Center for Strategic and International Studies (CSIS), where she concurrently serves as executive director of the China Balance Sheet project. She was a fellow in the CSIS International Security Program, and has more than a decade of government and private sector experience working on issues relating to China.

Derek Mitchell is a senior fellow for Asia at the Center for Strategic and International Studies in Washington, D.C. Mr. Mitchell was Special Assistant for Asian and Pacific Affairs, Office of the Secretary of Defense, from 1997 to 2001, when he served alternately as Senior Country Director for China, Taiwan, Mongolia, and Hong Kong, Director for Regional Security Affairs, Country Director for Japan, and Senior Country Director for the Philippines, Indonesia, Malaysia, Brunei, and Singapore. Mr. Mitchell has written widely on Asian and U.S.-Asian affairs, and is the coauthor of *China: The Balance Sheet—What the World Needs to Know Now About the Emerging Superpower* (2006).

Matthew Oresman is a senior advisor with the China-Eurasia Forum (www.chinaeurasia.org). Prior to founding the China-Eurasia Forum, Mr. Oresman was a researcher with the Freeman Chair in China Studies at the Center for Strategic and International Studies (CSIS), and coordinator for the Freeman Chair's project on China's emergence in Central Asia. As such, he was the principal author of *China's New Journey to the West: Report on China's*

Emergence in Central Asia and Implications for U.S. Interests (August 2003). Mr. Oresman has published widely on China-Central Asia relations and has given presentations at Chinese and American think tanks and government departments on this topic. He received a B.A. in international relations from the University of Pennsylvania in 2002. He studied at Fudan University in Shanghai in 2000, and at Tsinghua University in Beijing from 2003 to 2004. He has worked previously at the State Department, in the U.S. Senate, and in the U.S. Department of Justice. In addition, Mr. Oresman is currently a J.D. candidate at Georgetown University Law Center and will be joining an international law firm in Washington, DC next year.

Chung-chian Teng received his M.A. and Ph.D. in political science at Northwestern University in 1981 and 1985, respectively. Currently, he is a professor in the Department of Diplomacy at National Chengchi University (Taipei, Taiwan). His most recent coauthored books include *Latin American Studies* (2002, in Chinese), *Political Communication and Skills of Negotiation* (2003, in Chinese), and *Governments and Politics* (2003, in Chinese).

Index